Mometrix
TEST PREPARATION

Common Core High School: Functions
Secrets Study Guide

I0802514

DEAR FUTURE EXAM SUCCESS STORY

First of all, **THANK YOU** for purchasing Mometrix study materials!

Second, congratulations! You are one of the few determined test-takers who are committed to doing whatever it takes to excel on your exam. **You have come to the right place.** We developed these study materials with one goal in mind: to deliver you the information you need in a format that's concise and easy to use.

In addition to optimizing your guide for the content of the test, we've outlined our recommended steps for breaking down the preparation process into small, attainable goals so you can make sure you stay on track.

We've also analyzed the entire test-taking process, identifying the most common pitfalls and showing how you can overcome them and be ready for any curveball the test throws you.

Standardized testing is one of the biggest obstacles on your road to success, which only increases the importance of doing well in the high-pressure, high-stakes environment of test day. Your results on this test could have a significant impact on your future, and this guide provides the information and practical advice to help you achieve your full potential on test day.

Your success is our success

We would love to hear from you! If you would like to share the story of your exam success or if you have any questions or comments in regard to our products, please contact us at **800-673-8175** or **support@mometrix.com**.

Thanks again for your business and we wish you continued success!

Sincerely,
The Mometrix Test Preparation Team

Need more help? Check out our flashcards at:
http://MometrixFlashcards.com/CCSS

Written and edited by the Mometrix Exam Secrets Test Prep Team
Printed in the United States of America

TABLE OF CONTENTS

Introduction

Thank you for purchasing this resource! You have made the choice to prepare yourself for a test that could have a huge impact on your future, and this guide is designed to help you be fully ready for test day. Obviously, it's important to have a solid understanding of the test material, but you also need to be prepared for the unique environment and stressors of the test, so that you can perform to the best of your abilities.

For this purpose, the first section that appears in this guide is the **Secret Keys**. We've devoted countless hours to meticulously researching what works and what doesn't, and we've boiled down our findings to the five most impactful steps you can take to improve your performance on the test. We start at the beginning with study planning and move through the preparation process, all the way to the testing strategies that will help you get the most out of what you know when you're finally sitting in front of the test.

We recommend that you start preparing for your test as far in advance as possible. However, if you've bought this guide as a last-minute study resource and only have a few days before your test, we recommend that you skip over the first two Secret Keys since they address a long-term study plan.

If you struggle with **test anxiety**, we strongly encourage you to check out our recommendations for how you can overcome it. Test anxiety is a formidable foe, but it can be beaten, and we want to make sure you have the tools you need to defeat it.

Secret Key #1 – Plan Big, Study Small

There's a lot riding on your performance. If you want to ace this test, you're going to need to keep your skills sharp and the material fresh in your mind. You need a plan that lets you review everything you need to know while still fitting in your schedule. We'll break this strategy down into three categories.

Information Organization

Start with the information you already have: the official test outline. From this, you can make a complete list of all the concepts you need to cover before the test. Organize these concepts into groups that can be studied together, and create a list of any related vocabulary you need to learn so you can brush up on any difficult terms. You'll want to keep this vocabulary list handy once you actually start studying since you may need to add to it along the way.

Time Management

Once you have your set of study concepts, decide how to spread them out over the time you have left before the test. Break your study plan into small, clear goals so you have a manageable task for each day and know exactly what you're doing. Then just focus on one small step at a time. When you manage your time this way, you don't need to spend hours at a time studying. Studying a small block of content for a short period each day helps you retain information better and avoid stressing over how much you have left to do. You can relax knowing that you have a plan to cover everything in time. In order for this strategy to be effective though, you have to start studying early and stick to your schedule. Avoid the exhaustion and futility that comes from last-minute cramming!

Study Environment

The environment you study in has a big impact on your learning. Studying in a coffee shop, while probably more enjoyable, is not likely to be as fruitful as studying in a quiet room. It's important to keep distractions to a minimum. You're only planning to study for a short block of time, so make the most of it. Don't pause to check your phone or get up to find a snack. It's also important to **avoid multitasking**. Research has consistently shown that multitasking will make your studying dramatically less effective. Your study area should also be comfortable and well-lit so you don't have the distraction of straining your eyes or sitting on an uncomfortable chair.

The time of day you study is also important. You want to be rested and alert. Don't wait until just before bedtime. Study when you'll be most likely to comprehend and remember. Even better, if you know what time of day your test will be, set that time aside for study. That way your brain will be used to working on that subject at that specific time and you'll have a better chance of recalling information.

Finally, it can be helpful to team up with others who are studying for the same test. Your actual studying should be done in as isolated an environment as possible, but the work of organizing the information and setting up the study plan can be divided up. In between study sessions, you can discuss with your teammates the concepts that you're all studying and quiz each other on the details. Just be sure that your teammates are as serious about the test as you are. If you find that your study time is being replaced with social time, you might need to find a new team.

Secret Key #2 – Make Your Studying Count

You're devoting a lot of time and effort to preparing for this test, so you want to be absolutely certain it will pay off. This means doing more than just reading the content and hoping you can remember it on test day. It's important to make every minute of study count. There are two main areas you can focus on to make your studying count.

Retention

It doesn't matter how much time you study if you can't remember the material. You need to make sure you are retaining the concepts. To check your retention of the information you're learning, try recalling it at later times with minimal prompting. Try carrying around flashcards and glance at one or two from time to time or ask a friend who's also studying for the test to quiz you.

To enhance your retention, look for ways to put the information into practice so that you can apply it rather than simply recalling it. If you're using the information in practical ways, it will be much easier to remember. Similarly, it helps to solidify a concept in your mind if you're not only reading it to yourself but also explaining it to someone else. Ask a friend to let you teach them about a concept you're a little shaky on (or speak aloud to an imaginary audience if necessary). As you try to summarize, define, give examples, and answer your friend's questions, you'll understand the concepts better and they will stay with you longer. Finally, step back for a big picture view and ask yourself how each piece of information fits with the whole subject. When you link the different concepts together and see them working together as a whole, it's easier to remember the individual components.

Finally, practice showing your work on any multi-step problems, even if you're just studying. Writing out each step you take to solve a problem will help solidify the process in your mind, and you'll be more likely to remember it during the test.

Modality

Modality simply refers to the means or method by which you study. Choosing a study modality that fits your own individual learning style is crucial. No two people learn best in exactly the same way, so it's important to know your strengths and use them to your advantage.

For example, if you learn best by visualization, focus on visualizing a concept in your mind and draw an image or a diagram. Try color-coding your notes, illustrating them, or creating symbols that will trigger your mind to recall a learned concept. If you learn best by hearing or discussing information, find a study partner who learns the same way or read aloud to yourself. Think about how to put the information in your own words. Imagine that you are giving a lecture on the topic and record yourself so you can listen to it later.

For any learning style, flashcards can be helpful. Organize the information so you can take advantage of spare moments to review. Underline key words or phrases. Use different colors for different categories. Mnemonic devices (such as creating a short list in which every item starts with the same letter) can also help with retention. Find what works best for you and use it to store the information in your mind most effectively and easily.

Secret Key #3 – Practice the Right Way

Your success on test day depends not only on how many hours you put into preparing, but also on whether you prepared the right way. It's good to check along the way to see if your studying is paying off. One of the most effective ways to do this is by taking practice tests to evaluate your progress. Practice tests are useful because they show exactly where you need to improve. Every time you take a practice test, pay special attention to these three groups of questions:

- The questions you got wrong
- The questions you had to guess on, even if you guessed right
- The questions you found difficult or slow to work through

This will show you exactly what your weak areas are, and where you need to devote more study time. Ask yourself why each of these questions gave you trouble. Was it because you didn't understand the material? Was it because you didn't remember the vocabulary? Do you need more repetitions on this type of question to build speed and confidence? Dig into those questions and figure out how you can strengthen your weak areas as you go back to review the material.

Additionally, many practice tests have a section explaining the answer choices. It can be tempting to read the explanation and think that you now have a good understanding of the concept. However, an explanation likely only covers part of the question's broader context. Even if the explanation makes perfect sense, **go back and investigate** every concept related to the question until you're positive you have a thorough understanding.

As you go along, keep in mind that the practice test is just that: practice. Memorizing these questions and answers will not be very helpful on the actual test because it is unlikely to have any of the same exact questions. If you only know the right answers to the sample questions, you won't be prepared for the real thing. **Study the concepts** until you understand them fully, and then you'll be able to answer any question that shows up on the test.

It's important to wait on the practice tests until you're ready. If you take a test on your first day of study, you may be overwhelmed by the amount of material covered and how much you need to learn. Work up to it gradually.

On test day, you'll need to be prepared for answering questions, managing your time, and using the test-taking strategies you've learned. It's a lot to balance, like a mental marathon that will have a big impact on your future. Like training for a marathon, you'll need to start slowly and work your way up. When test day arrives, you'll be ready.

Start with the strategies you've read in the first two Secret Keys—plan your course and study in the way that works best for you. If you have time, consider using multiple study resources to get different approaches to the same concepts. It can be helpful to see difficult concepts from more than one angle. Then find a good source for practice tests. Many times, the test website will suggest potential study resources or provide sample tests.

Practice Test Strategy

If you're able to find at least three practice tests, we recommend this strategy:

Untimed and Open-Book Practice

Take the first test with no time constraints and with your notes and study guide handy. Take your time and focus on applying the strategies you've learned.

Timed and Open-Book Practice

Take the second practice test open-book as well, but set a timer and practice pacing yourself to finish in time.

Timed and Closed-Book Practice

Take any other practice tests as if it were test day. Set a timer and put away your study materials. Sit at a table or desk in a quiet room, imagine yourself at the testing center, and answer questions as quickly and accurately as possible.

Keep repeating timed and closed-book tests on a regular basis until you run out of practice tests or it's time for the actual test. Your mind will be ready for the schedule and stress of test day, and you'll be able to focus on recalling the material you've learned.

Secret Key #4 – Pace Yourself

Once you're fully prepared for the material on the test, your biggest challenge on test day will be managing your time. Just knowing that the clock is ticking can make you panic even if you have plenty of time left. Work on pacing yourself so you can build confidence against the time constraints of the exam. Pacing is a difficult skill to master, especially in a high-pressure environment, so **practice is vital**.

Set time expectations for your pace based on how much time is available. For example, if a section has 60 questions and the time limit is 30 minutes, you know you have to average 30 seconds or less per question in order to answer them all. Although 30 seconds is the hard limit, set 25 seconds per question as your goal, so you reserve extra time to spend on harder questions. When you budget extra time for the harder questions, you no longer have any reason to stress when those questions take longer to answer.

Don't let this time expectation distract you from working through the test at a calm, steady pace, but keep it in mind so you don't spend too much time on any one question. Recognize that taking extra time on one question you don't understand may keep you from answering two that you do understand later in the test. If your time limit for a question is up and you're still not sure of the answer, mark it and move on, and come back to it later if the time and the test format allow. If the testing format doesn't allow you to return to earlier questions, just make an educated guess; then put it out of your mind and move on.

On the easier questions, be careful not to rush. It may seem wise to hurry through them so you have more time for the challenging ones, but it's not worth missing one if you know the concept and just didn't take the time to read the question fully. Work efficiently but make sure you understand the question and have looked at all of the answer choices, since more than one may seem right at first.

Even if you're paying attention to the time, you may find yourself a little behind at some point. You should speed up to get back on track, but do so wisely. Don't panic; just take a few seconds less on each question until you're caught up. Don't guess without thinking, but do look through the answer choices and eliminate any you know are wrong. If you can get down to two choices, it is often worthwhile to guess from those. Once you've chosen an answer, move on and don't dwell on any that you skipped or had to hurry through. If a question was taking too long, chances are it was one of the harder ones, so you weren't as likely to get it right anyway.

On the other hand, if you find yourself getting ahead of schedule, it may be beneficial to slow down a little. The more quickly you work, the more likely you are to make a careless mistake that will affect your score. You've budgeted time for each question, so don't be afraid to spend that time. Practice an efficient but careful pace to get the most out of the time you have.

Secret Key #5 – Have a Plan for Guessing

When you're taking the test, you may find yourself stuck on a question. Some of the answer choices seem better than others, but you don't see the one answer choice that is obviously correct. What do you do?

The scenario described above is very common, yet most test takers have not effectively prepared for it. Developing and practicing a plan for guessing may be one of the single most effective uses of your time as you get ready for the exam.

In developing your plan for guessing, there are three questions to address:

- When should you start the guessing process?
- How should you narrow down the choices?
- Which answer should you choose?

When to Start the Guessing Process

Unless your plan for guessing is to select C every time (which, despite its merits, is not what we recommend), you need to leave yourself enough time to apply your answer elimination strategies. Since you have a limited amount of time for each question, that means that if you're going to give yourself the best shot at guessing correctly, you have to decide quickly whether or not you will guess.

Of course, the best-case scenario is that you don't have to guess at all, so first, see if you can answer the question based on your knowledge of the subject and basic reasoning skills. Focus on the key words in the question and try to jog your memory of related topics. Give yourself a chance to bring the knowledge to mind, but once you realize that you don't have (or you can't access) the knowledge you need to answer the question, it's time to start the guessing process.

It's almost always better to start the guessing process too early than too late. It only takes a few seconds to remember something and answer the question from knowledge. Carefully eliminating wrong answer choices takes longer. Plus, going through the process of eliminating answer choices can actually help jog your memory.

Summary: Start the guessing process as soon as you decide that you can't answer the question based on your knowledge.

How to Narrow Down the Choices

The next chapter in this book (**Test-Taking Strategies**) includes a wide range of strategies for how to approach questions and how to look for answer choices to eliminate. You will definitely want to read those carefully, practice them, and figure out which ones work best for you. Here though, we're going to address a mindset rather than a particular strategy.

Your odds of guessing an answer correctly depend on how many options you are choosing from.

Number of options left	5	4	3	2	1
Odds of guessing correctly	20%	25%	33%	50%	100%

You can see from this chart just how valuable it is to be able to eliminate incorrect answers and make an educated guess, but there are two things that many test takers do that cause them to miss out on the benefits of guessing:

- Accidentally eliminating the correct answer
- Selecting an answer based on an impression

We'll look at the first one here, and the second one in the next section.

To avoid accidentally eliminating the correct answer, we recommend a thought exercise called **the $5 challenge**. In this challenge, you only eliminate an answer choice from contention if you are willing to bet $5 on it being wrong. Why $5? Five dollars is a small but not insignificant amount of money. It's an amount you could afford to lose but wouldn't want to throw away. And while losing $5 once might not hurt too much, doing it twenty times will set you back $100. In the same way, each small decision you make—eliminating a choice here, guessing on a question there—won't by itself impact your score very much, but when you put them all together, they can make a big difference. By holding each answer choice elimination decision to a higher standard, you can reduce the risk of accidentally eliminating the correct answer.

The $5 challenge can also be applied in a positive sense: If you are willing to bet $5 that an answer choice *is* correct, go ahead and mark it as correct.

Summary: Only eliminate an answer choice if you are willing to bet $5 that it is wrong.

Which Answer to Choose

You're taking the test. You've run into a hard question and decided you'll have to guess. You've eliminated all the answer choices you're willing to bet $5 on. Now you have to pick an answer. Why do we even need to talk about this? Why can't you just pick whichever one you feel like when the time comes?

The answer to these questions is that if you don't come into the test with a plan, you'll rely on your impression to select an answer choice, and if you do that, you risk falling into a trap. The test writers know that everyone who takes their test will be guessing on some of the questions, so they intentionally write wrong answer choices to seem plausible. You still have to pick an answer though, and if the wrong answer choices are designed to look right, how can you ever be sure that you're not falling for their trap? The best solution we've found to this dilemma is to take the decision out of your hands entirely. Here is the process we recommend:

Once you've eliminated any choices that you are confident (willing to bet $5) are wrong, select the first remaining choice as your answer.

Whether you choose to select the first remaining choice, the second, or the last, the important thing is that you use some preselected standard. Using this approach guarantees that you will not be enticed into selecting an answer choice that looks right, because you are not basing your decision on how the answer choices look.

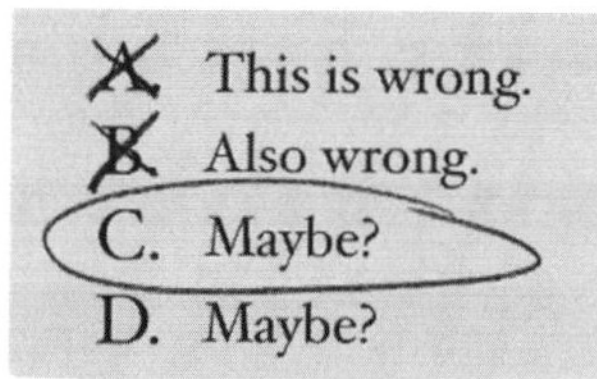

This is not meant to make you question your knowledge. Instead, it is to help you recognize the difference between your knowledge and your impressions. There's a huge difference between thinking an answer is right because of what you know, and thinking an answer is right because it looks or sounds like it should be right.

Summary: To ensure that your selection is appropriately random, make a predetermined selection from among all answer choices you have not eliminated.

Test-Taking Strategies

This section contains a list of test-taking strategies that you may find helpful as you work through the test. By taking what you know and applying logical thought, you can maximize your chances of answering any question correctly!

It is very important to realize that every question is different and every person is different: no single strategy will work on every question, and no single strategy will work for every person. That's why we've included all of them here, so you can try them out and determine which ones work best for different types of questions and which ones work best for you.

Question Strategies

☑ READ CAREFULLY

Read the question and the answer choices carefully. Don't miss the question because you misread the terms. You have plenty of time to read each question thoroughly and make sure you understand what is being asked. Yet a happy medium must be attained, so don't waste too much time. You must read carefully and efficiently.

☑ CONTEXTUAL CLUES

Look for contextual clues. If the question includes a word you are not familiar with, look at the immediate context for some indication of what the word might mean. Contextual clues can often give you all the information you need to decipher the meaning of an unfamiliar word. Even if you can't determine the meaning, you may be able to narrow down the possibilities enough to make a solid guess at the answer to the question.

☑ PREFIXES

If you're having trouble with a word in the question or answer choices, try dissecting it. Take advantage of every clue that the word might include. Prefixes can be a huge help. Usually, they allow you to determine a basic meaning. *Pre-* means before, *post-* means after, *pro-* is positive, *de-* is negative. From prefixes, you can get an idea of the general meaning of the word and try to put it into context.

☑ HEDGE WORDS

Watch out for critical hedge words, such as *likely, may, can, often, almost, mostly, usually, generally, rarely,* and *sometimes.* Question writers insert these hedge phrases to cover every possibility. Often an answer choice will be wrong simply because it leaves no room for exception. Be on guard for answer choices that have definitive words such as *exactly* and *always.*

☑ SWITCHBACK WORDS

Stay alert for *switchbacks.* These are the words and phrases frequently used to alert you to shifts in thought. The most common switchback words are *but, although,* and *however.* Others include *nevertheless, on the other hand, even though, while, in spite of, despite,* and *regardless of.* Switchback words are important to catch because they can change the direction of the question or an answer choice.

☑ Face Value

When in doubt, use common sense. Accept the situation in the problem at face value. Don't read too much into it. These problems will not require you to make wild assumptions. If you have to go beyond creativity and warp time or space in order to have an answer choice fit the question, then you should move on and consider the other answer choices. These are normal problems rooted in reality. The applicable relationship or explanation may not be readily apparent, but it is there for you to figure out. Use your common sense to interpret anything that isn't clear.

Answer Choice Strategies

☑ Answer Selection

The most thorough way to pick an answer choice is to identify and eliminate wrong answers until only one is left, then confirm it is the correct answer. Sometimes an answer choice may immediately seem right, but be careful. The test writers will usually put more than one reasonable answer choice on each question, so take a second to read all of them and make sure that the other choices are not equally obvious. As long as you have time left, it is better to read every answer choice than to pick the first one that looks right without checking the others.

☑ Answer Choice Families

An answer choice family consists of two (in rare cases, three) answer choices that are very similar in construction and cannot all be true at the same time. If you see two answer choices that are direct opposites or parallels, one of them is usually the correct answer. For instance, if one answer choice says that quantity *x* increases and another either says that quantity *x* decreases (opposite) or says that quantity *y* increases (parallel), then those answer choices would fall into the same family. An answer choice that doesn't match the construction of the answer choice family is more likely to be incorrect. Most questions will not have answer choice families, but when they do appear, you should be prepared to recognize them.

☑ Eliminate Answers

Eliminate answer choices as soon as you realize they are wrong, but make sure you consider all possibilities. If you are eliminating answer choices and realize that the last one you are left with is also wrong, don't panic. Start over and consider each choice again. There may be something you missed the first time that you will realize on the second pass.

☑ Avoid Fact Traps

Don't be distracted by an answer choice that is factually true but doesn't answer the question. You are looking for the choice that answers the question. Stay focused on what the question is asking for so you don't accidentally pick an answer that is true but incorrect. Always go back to the question and make sure the answer choice you've selected actually answers the question and is not merely a true statement.

☑ Extreme Statements

In general, you should avoid answers that put forth extreme actions as standard practice or proclaim controversial ideas as established fact. An answer choice that states the "process should be used in certain situations, if..." is much more likely to be correct than one that states the "process should be discontinued completely." The first is a calm rational statement and doesn't even make a definitive, uncompromising stance, using a hedge word *if* to provide wiggle room, whereas the second choice is far more extreme.

⊘ Benchmark

As you read through the answer choices and you come across one that seems to answer the question well, mentally select that answer choice. This is not your final answer, but it's the one that will help you evaluate the other answer choices. The one that you selected is your benchmark or standard for judging each of the other answer choices. Every other answer choice must be compared to your benchmark. That choice is correct until proven otherwise by another answer choice beating it. If you find a better answer, then that one becomes your new benchmark. Once you've decided that no other choice answers the question as well as your benchmark, you have your final answer.

⊘ Predict the Answer

Before you even start looking at the answer choices, it is often best to try to predict the answer. When you come up with the answer on your own, it is easier to avoid distractions and traps because you will know exactly what to look for. The right answer choice is unlikely to be word-for-word what you came up with, but it should be a close match. Even if you are confident that you have the right answer, you should still take the time to read each option before moving on.

General Strategies

⊘ Tough Questions

If you are stumped on a problem or it appears too hard or too difficult, don't waste time. Move on! Remember though, if you can quickly check for obviously incorrect answer choices, your chances of guessing correctly are greatly improved. Before you completely give up, at least try to knock out a couple of possible answers. Eliminate what you can and then guess at the remaining answer choices before moving on.

⊘ Check Your Work

Since you will probably not know every term listed and the answer to every question, it is important that you get credit for the ones that you do know. Don't miss any questions through careless mistakes. If at all possible, try to take a second to look back over your answer selection and make sure you've selected the correct answer choice and haven't made a costly careless mistake (such as marking an answer choice that you didn't mean to mark). This quick double check should more than pay for itself in caught mistakes for the time it costs.

⊘ Pace Yourself

It's easy to be overwhelmed when you're looking at a page full of questions; your mind is confused and full of random thoughts, and the clock is ticking down faster than you would like. Calm down and maintain the pace that you have set for yourself. Especially as you get down to the last few minutes of the test, don't let the small numbers on the clock make you panic. As long as you are on track by monitoring your pace, you are guaranteed to have time for each question.

⊘ Don't Rush

It is very easy to make errors when you are in a hurry. Maintaining a fast pace in answering questions is pointless if it makes you miss questions that you would have gotten right otherwise. Test writers like to include distracting information and wrong answers that seem right. Taking a little extra time to avoid careless mistakes can make all the difference in your test score. Find a pace that allows you to be confident in the answers that you select.

✓ KEEP MOVING

Panicking will not help you pass the test, so do your best to stay calm and keep moving. Taking deep breaths and going through the answer elimination steps you practiced can help to break through a stress barrier and keep your pace.

Final Notes

The combination of a solid foundation of content knowledge and the confidence that comes from practicing your plan for applying that knowledge is the key to maximizing your performance on test day. As your foundation of content knowledge is built up and strengthened, you'll find that the strategies included in this chapter become more and more effective in helping you quickly sift through the distractions and traps of the test to isolate the correct answer.

Now that you're preparing to move forward into the test content chapters of this book, be sure to keep your goal in mind. As you read, think about how you will be able to apply this information on the test. If you've already seen sample questions for the test and you have an idea of the question format and style, try to come up with questions of your own that you can answer based on what you're reading. This will give you valuable practice applying your knowledge in the same ways you can expect to on test day.

Good luck and good studying!

Functions

A function is a relation between two sets, the *domain* and the *range*, such that each element of the domain is associated with exactly one element of the range.

There are other ways this definition can be phrased. A less technical wording is that a function is a relation that associates each possible value of one variable, x, with a unique value of a second variable, y.

Note that it is *not* necessarily the case that each element of the *range* is associated with exactly one element of the *domain* (or, under the other definition, that each value of y is associated with a unique value of x). However, a relation for which this condition does hold is *invertible*—in this case, interchanging the domain and the range of the function gives rise to another function (the original function's *inverse*).

LINEAR, QUADRATIC, POLYNOMIAL, AND RATIONAL FUNCTIONS

A *linear* function is a function the graph of which is a straight line. The expression for a linear function includes x (possibly multiplied by some constant coefficient, and possibly added to some other constant), but does not include x^2 or any higher powers of x. The following, for example, are all linear functions: $f(x) = x$, $f(x) = 1 - 2x$, $f(x) = \frac{3}{4}x + 3$.

The expression for a *quadratic* function includes a term containing x^2 (again possibly multiplied by some constant coefficient), but no term contains x^3 or any higher powers of x. The following, for example, are all quadratic functions: $f(x) = x^2 + x + 1$, $f(x) = 3x^2 - 2$, $f(x) = -\frac{1}{2}x^2$.

The expression for a *polynomial* function is a sum of one or more powers of x, possibly multiplied by constant coefficients. (Note that $1 = x^0$ is a power of x.) All linear and quadratic functions are polynomials; so are the following: $f(x) = 3x^7$, $f(x) = 2x^3 - 2x^2 + 5x - 6$, $f(x) = x^{2000} - x^{169}$.

A rational function is a ratio of polynomials. Examples include $\frac{x^2-1}{x}$, $\frac{1}{3x^3+2x^2+x}$, and $\frac{x^4+3}{x^2-2}$.

DETERMINE HOW YOU CAN TELL FROM LOOKING AT A GRAPH WHETHER THE GRAPH REPRESENTS A FUNCTION

The simplest way to tell by visual inspection whether a graph represents a function is a procedure sometimes known as the "vertical line test". Imagine a vertical line scanning across the graph. If, at any point, the line intersects more than one point on the graph, the graph does *not* represent a function. Otherwise, it does.

For example, consider the following graphs.

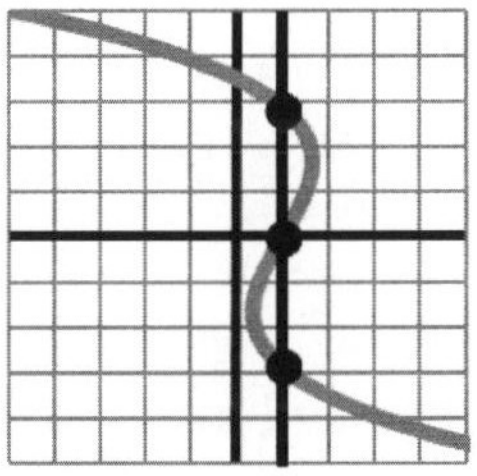 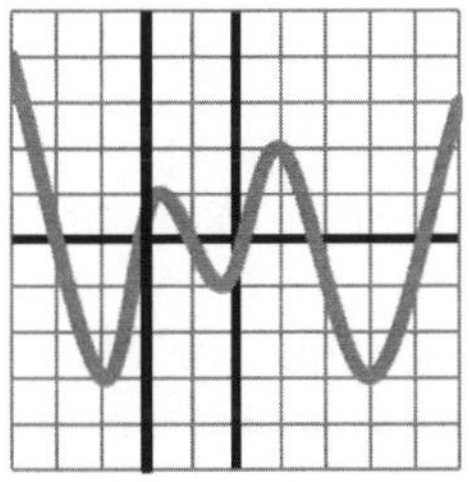

The vertical line in the image on the left intersects the graph in three points. Therefore, the graph on the left does *not* represent a function. On the right, however, no matter where we place the vertical line, it will never intersect more than one point. Therefore, the graph on the right represents a function.

FINDING THE DOMAIN OF THE FUNCTION

Given a function expressed in a table like the one below, determine how you can find the domain of the function.

x	0	3	2	1	8	10	6	7
$f(x)$	2	8	6	8	2	3	4	5

The *domain* of a function $f(x)$ is the set of all input values for which the function is defined. For a function expressed in a table like the one in this example, it's simple to find the domain, because every point in the function—that is, every input-output pair—is given explicitly. To find the domain, we just list all the x values: in this case, that would be {0, 3, 2, 1, 8, 10, 6, 7}, or, putting them in ascending order, {0, 1, 2, 3, 6, 7, 8, 10}. (Putting the values in ascending order isn't strictly necessary, but generally makes the set easier to read.)

Note that we don't have to worry about the possibility of an input value being repeated: by definition of a function, no input value can be matched to more than one output value.

Given a function expressed in analytic form such as $f(x) = \sin x$, $g(x) = \sqrt{x+1}$, or $h(x) = \frac{x}{x^2-1}$, explain how you can find the domain of the function.

The *domain* of a function $f(x)$ is the set of all possible input values of the function—that is, of every possible value of x for which the function is defined. Some functions, such as $f(x) = \sin x$, are defined for any real value of x; in this case, the domain of $f(x)$ is $(-\infty, \infty)$, that is, it includes all real numbers. There are functions, however, for which that isn't true. For instance, consider $g(x) = \sqrt{x+1}$. The square root of a negative number is not defined (at least as a real number), so in order for $g(x)$ to be defined we must have $x + 1 \geq 0$, so $x \geq -1$. The domain of $g(x)$ is thus $[-1, \infty)$. Similarly, $h(x) = \frac{x}{x^2-1}$ is not defined when the denominator is zero, which is true when $x = \pm 1$. The domain therefore includes all real numbers *except* 1 and -1, which we can write as $(-\infty, -1) \cup (-1,1) \cup (1, \infty)$.

FINDING THE RANGE OF A FUNCTION

GIVEN A FUNCTION EXPRESSED IN A TABLE LIKE THE ONE BELOW, DETERMINE HOW YOU CAN FIND THE RANGE OF THE FUNCTION.

x	-1	4	2	1	0	3	8	6
$f(x)$	3	0	3	−1	−1	2	4	6

The *range* of a function $f(x)$ is the set of all possible output values of the function—that is, of every possible value of $f(x)$, for any value of x in the function's domain. For a function expressed in a table like the one in this example, it's simple to find the range, because every point in the function—that is, every input-output pair—is given explicitly. To find the range, we just list all the values of $f(x)$; in this case, that would be {3, 0, 3, -1, -1, 2, 4, 6}. Note that some of these values appear more than once. This is entirely permissible for a function; while each value of x must be matched to a unique value of $f(x)$, the converse is not true. We don't need to list each value more than once, so eliminating duplicates, the range is {3, 0, -1, 2, 4, 6}, or, putting them in ascending order, {-1, 0, 2, 3, 4, 6}. (Putting the values in ascending order isn't strictly necessary, but generally makes the set easier to read.)

Given a function expressed in analytic form such as $f(x) = x^3$, $g(x) = x^2 - 2x + 4$, or $h(x) = 1/x$, explain how you can find the range of the function.

For a continuous function, we can find the range by just finding the maximum and minimum values of the function; the range comprises all the points in between. Some functions have no maximum or minimum; this is the case, for instance, with $f(x) = x^3$. At the left, the function decreases indefinitely, and on the right, it increases indefinitely. The range therefore includes all real numbers, $(-\infty, \infty)$. Other functions may have a maximum, but no minimum, or vice versa. The quadratic function $g(x) = x^2 - 2x + 4$ has a minimum at $(1,3)$, but no maximum, so its range is $[3, \infty)$.

For a discontinuous function, we must examine each part separately. $h(x) = 1/x$ has a discontinuity at $x = 0$. On the left side, the function has no minimum, but has an asymptotic maximum of 0 (asymptotic because it approaches 0 but never actually reaches it); the range of this side is $(-\infty, 0)$. Similarly, the right side has a range of $(0, \infty)$. The total range of the function is the union of these two ranges, hence $(-\infty, 0) \cup (0, \infty)$.

Example

Suppose $\mathrm{f(x)} = \mathrm{x}^2 + 1$ and $\mathrm{g(x)} = \mathrm{x} - 1$. How would you find each of the following?

$f(3)$ $\quad$ $g(3)$ $\quad$ $f(y^2)$ $\quad$ $g(z + 1)$

In function notation, "$f(3)$" can be thought of as a shorthand for "$f(x)$ when $x = 3$". To evaluate the function at this point, just replace the x in the expression for the function with the number. If $f(x) = x^2 + 1$, then to find $f(3)$ we just replace each x in "$x^2 + 1$" with 3. So $f(3) = 3^2 + 1$, which we can simplify to 10. Similarly, $g(3) = 3 - 1 = 2$.

The same principle holds when instead of a number, there is a variable or an expression within the parentheses. To find $f(y^2)$, where $f(x) = x^2 + 1$, we replace each x in "$x^2 + 1$" with "y^2" (or "(y^2)"—including parentheses may make it easier to keep track of the substitution). This yields $f(y^2) = (y^2)^2 + 1 = y^4 + 1$. Similarly,

$g(z + 1) = (z + 1) - 1 = z$.

Example

Describe in words what a statement such as $\mathrm{f(x)} = \mathrm{x}^2 + 1$ means

This kind of statement defines a particular function; it tells how to match up each point in the domain with a point in the range. The possible values of x comprise the domain—this can usually be assumed to include all real numbers unless specified otherwise, or unless there are some points for which the expression on the right side of the equation is undefined. Each value of x is then matched with a (not necessarily unique) value of $f(x)$ (or y); all the values of $f(x)$ that are matched to values of x comprise the range.

In the case of the statement $f(x) = x^2 + 1$, this defines a function in which each real number x is matched to the number $x^2 + 1$. For instance, $x = 0$ corresponds to $f(0) = 0^2 + 1 = 1$, when $x = 1$ $f(1) = 1^2 + 1 = 2$, and so forth. Note that $f(x)$ in this case can never be less than 1, so the range of this function is $[1, \infty)$.

Real-World Situations That Can Be Modeled by Statements That Use Function Notation

A statement using function notation can model any situation in which one quantity depends uniquely on one or more other quantities. For example, the area of a rectangle can be expressed as a function of its width and height. The maximum vertical distance a projectile travels can be expressed as a function of its initial vertical speed. An object's position, the amount of money in a bank account, or any other quantity that changes over time can be expressed as a function of time.

A relationship *cannot* be modeled with a function, however, if it involves two quantities neither of which is uniquely determined by the other—that is, if each quantity may have multiple values corresponding to the same value of the other. For example, we could *not* write a function to represent the relationship between peoples' height in inches and their weight in pounds. There are people of the same height with different weights, and people of the same weight but different heights.

Defining a Function Recursively

To define a function recursively is to give the value of the function for a given input value in terms of the function's values for other input values—for example, to define $f(x)$ in terms of $f(x-1)$. A function cannot be defined *purely* recursively; it's necessary to define the function absolutely for at least one point to give a starting point for the recursion.

For example, the factorial function, $f(n) = n! = n(n-1)(n-2) \ldots (2)(1)$, can be defined recursively as follows: $f(1) = 1$, otherwise $f(n) = n \cdot f(n-1)$. The domain of this function is the positive integers. (Technically, the factorial function is also defined at $n = 0$, where $0! = 1$, but that's not important for this example.) Then, for example, $f(3) = 3 \cdot f(2) = 3 \cdot 2 \cdot f(1) = 3 \cdot 2 \cdot 1$.

Sequence

A sequence is a function in which the domain is confined to the integers, or some subset of the integers. Very often, the subset in question is the positive integers, in which case the sequence can be thought of as an ordered list of the elements of the range, with $f(n)$ identified directly with the nth element of the sequence. Such sequences are often written in list form; the sequence $f(n) = \frac{n^2+n}{2}$, where $f(1) = 1, f(2) = 3, f(3) = 6$, and so on, can simply be written as $1, 3, 6, 10, 15, \ldots$

Sequences are often defined recursively, with the first few elements of the sequence defined explicitly and with $f(n)$ otherwise defined in terms of $f(n-1)$ or in general of function values for smaller versions of n. Many functions can be equivalently defined explicitly or recursively. For instance, the sequence above, $f(n) = \frac{n^2+n}{2}$, can also be defined as $f(1) = 1$ and $f(n) = f(n-1) + n$.

FIBONACCI SEQUENCE

The Fibonacci sequence is a sequence in which the first two terms are 1 and each succeeding term is the sum of the preceding two terms: 1, 1, 2, 3, 5, 8, 13, 21... Written algebraically, $f(1) = f(2) = 1$, and thereafter $f(n) = f(n-1) + f(n-2)$. (There are some variations in its definition; sometimes the domain starts at 0 instead of 1, for instance, or the first term may be defined as 0 instead of 1.)

The Fibonacci sequence is the most common and perhaps the simplest sequence that is recursively defined in terms of more than one earlier term in the sequence (rather than each term being only defined in terms of the immediately preceding term). It also arises in many real-world phenomena.

Example

Given a function definition such as $(x) = x^2 - 9$, determine how you would find the function's *x* and *y* intercepts

To find the *y* intercept, simply find $f(0)$. The *y* intercept is by definition the point where the function crosses the *y*-axis, which means it's the function's value when $x = 0$. For the example function definition, $f(x) = x^2 - 9$, $f(0) = 0^2 - 9 = -9$. So the *y* intercept is (0, -9). Since no function can have multiple *y* values associated with a single value of *x*, a function cannot have more than one *y* intercept.

To find the *x* intercepts, we need to find where the function crosses the *x*-axis, so we have to find the value(s) of *x*, if any, for which $f(x) = 0$. There may be more than one such value. In the case of the example function, $f(x) = x^2 - 9$, this means $0 = x^2 - 9$. Solving for *x*, we get $x^2 = 9$, so $x = \pm 3$. So this function has two *x* intercepts, (-3, 0) and (3, 0).

ALL FUNCTIONS DO NOT NECESSARILY HAVE *X* AND *Y* INTERCEPTS

No, not all functions necessarily have *x* and *y* intercepts. To have a *y* intercept, a function must be defined at $x = 0$. This is not true of all functions: $f(x) = \sqrt{x-1}$, for instance, does not have a *y* intercept, because $f(0) = \sqrt{-1}$ is not defined in the real numbers. For a function $f(x)$ to have an *x* intercept, there must be one or more values of *x* such that $f(x) = 0$. This, again, is not true of all functions. For instance, for $f(x) = x^2 + 1$, there is no value of *x* for which $f(x) = 0$, since we cannot find a real value of *x* that satisfies the equation $x^2 + 1 = 0$. This function has no *x* intercept.

It is possible for a function to have neither an *x* nor a *y* intercept. One simple example is $f(x) = {}^1/_x$. Since ${}^1/_0$ is undefined, this function does not have a *y* intercept. Since there is no finite value of *x* satisfying ${}^1/_x = 0$, this function does not have an *x* intercept.

"RELATIVE MAXIMUM" AND "RELATIVE MINIMUM" OF A FUNCTION

A "relative maximum" of a function is a point at which the function has a higher value than any other point in its immediate vicinity. More technically, a relative maximum is a point (x, y) such that if we choose a sufficiently small interval around x, $f(x) > f(x')$ for any other point x' within the interval.

A relative minimum is just the opposite: a point at which the function has a *lower* value than any other point in its immediate vicinity, or, more technically, a point (x, y) such that if we choose a sufficiently small interval around x, $f(x) < f(x')$ for any other point x' within the interval.

A function may have multiple relative maxima and relative minima, or it may have none. A linear function such as $y = x$, for instance, has no relative maxima or minima, while $y = \sin(x)$ has infinitely many of each.

FINDING A RELATIVE MAXIMUM OR MINIMUM FROM A GRAPH

At a relative maximum, the graph goes from increasing to decreasing, forming a "peak" on the graph which is generally easy to spot visually. Similarly, at a relative minimum, the graph goes from decreasing to increasing, forming a "trough". On the graph below, for instance, points A and C are relative maxima, and point B is a relative minimum.

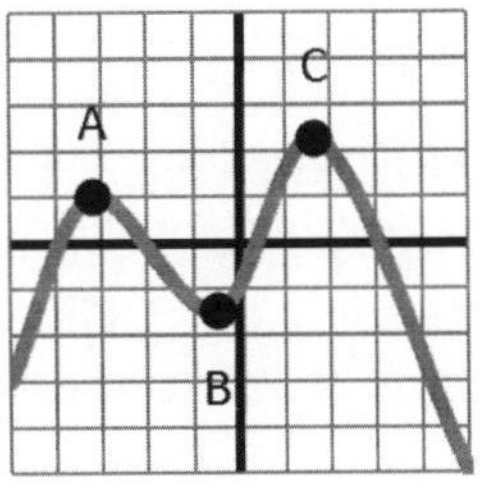

Note that point A is a relative maximum, but not an absolute maximum; the function has a higher value at point C. Similarly, point B is a relative minimum, but not an absolute minimum; the function has a lower value at the left and right ends of the graph than it does at point B. Point C, however, is both a relative and an absolute maximum; nowhere on the graph does the function have a higher value than at point C.

DETERMINING HOW YOU CAN TELL OVER WHAT INTERVALS A FUNCTION IS INCREASING OR DECREASING

If the function is presented in graph form, it's easy to tell by inspection where it's increasing and where it's decreasing. Where the slope of the graph is positive—where the graph goes from lower left to upper right—the function is increasing. Where the slope is negative—where the graph goes from upper left to lower right—the function is decreasing. In the graph below, for example, the graph is increasing over the intervals $(-\infty, -3)$ and $(0, 3)$, and decreasing over the intervals $(-3, 0)$ and $(3, \infty)$.

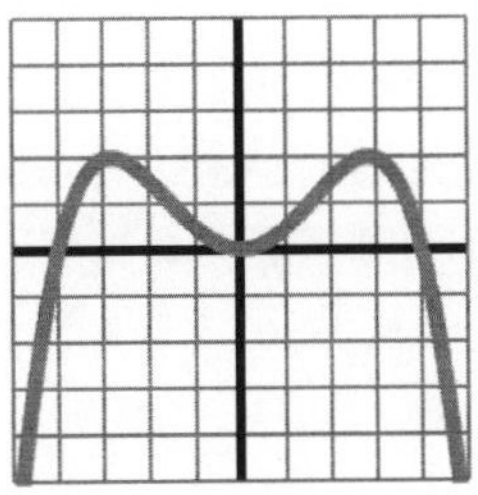

If the function is presented explicitly, such as $f(x) = x^2 - x^4$, we can graph the function and then find the appropriate intervals as above, but we can also, for instance, find the relative maxima and minima; an interval with a minimum on the left and a maximum on the right is increasing, and one with a maximum on the left and minimum on the right is decreasing.

If a Function Is Given Symbolically, Name Some Features We Can Find That May Be Useful in Sketching Its Graph

Sketching a graph without the aid of technology is facilitated by first determining some key properties of the function. If the function has horizontal asymptotes, we can use these as guidelines for the function's end behavior. If it doesn't have horizontal asymptotes, we can often determine the end behavior by other means. A function's vertical asymptotes likewise indicate its behavior near the corresponding values of x.

Also useful in determining the shape of a function are the relative maxima and minima, and the intervals over which the function is increasing or decreasing. The x and y intercepts provide a few points we know the function passes through. If the function has any symmetries or periodicity, these are also helpful in plotting it. Of course, if all else fails, we can also just pick a few x values at random, solve for $f(x)$ at those points, and plot those points to get a feel for the function's shape.

Periodic Functions

A function is *periodic* if it comprises successive repetitions of an identical shape. That is, a periodic function is unchanged if displaced by some constant P. This constant—the length of the repeating portion—is called the function's *period*. Periodicity can be defined in a more precise, mathematical form: a function $f(x)$ is periodic if $f(x + P) = f(x)$ for all x. (If this is true for P, it's also true for all multiples of P, but the function's period is defined as the smallest possible value of P for which the relation holds.)

For instance, the function $f(x) = \sin(x)$ is periodic with a period of 2π, since $\sin(x + 2\pi) = \sin(x)$ for all x. (It's also true that, for instance, $\sin(x + 4\pi) = \sin(x)$, and $\sin(x + 6\pi) = \sin(x)$, but there is no positive value of P *smaller* than 2π satisfying $\sin(x + P) = \sin(x)$.

Symmetrical Functions

A function is symmetrical if it remains unchanged under certain kinds of transformations. While there are many kinds of transformations and many kinds of symmetries, for functions the type of transformation most often considered is reflection. Particularly important are reflections through the y-axis and reflections through the origin.

A function is symmetrical under a reflection through the y-axis if it stays the same when "flipped over" around the y-axis into its mirror image—mathematically, $f(x) = f(-x)$. Such a function is by definition *even*. A function is symmetrical under a reflection through the origin if it stays the same when each point is reflected to the opposite side of the origin—mathematically, $f(x) = -f(-x)$. Such a function is by definition *odd*.

A periodic function is also symmetrical, not necessarily under reflection, but under an appropriate *translation*. The function is unchanged when translated (moved) horizontally by a distance equal to the function's period.

Function's End Behavior

A function's *end behavior* refers to its tendency at the extreme right and left sides of the graph—that is, what happens to the function as x tends toward $\pm\infty$. There are essentially three possibilities: either $f(x)$ increases without limit ($f(x)$ goes to ∞), $f(x)$ decreases without limit ($f(x)$ goes to

$-\infty$), or $f(x)$ tends toward some finite value (a horizontal asymptote). The behavior may be different at the two sides: while $f(x) = x^2$ goes to ∞ on both sides, $g(x) = x^3$ goes to ∞ on the right side and to $-\infty$ on the left, and $h(x) = e^{-x}$ goes to ∞ on the left and tends toward the horizontal asymptote $y = 0$ on the right.

WHEN WRITING A FUNCTION TO REPRESENT A QUANTITATIVE RELATIONSHIP GIVEN IN A WORD PROBLEM, EXPLAIN HOW TO DETERMINE THE FUNCTION'S DOMAIN

Sometimes the domain is given explicitly; if given the earnings of a company between 1980 and 2000, then if you write this as a function of the company's earnings over time in years, the domain would run from 1980 to 2000.

When not given explicitly, the domain must be deduced by considering logically within what limits the function applies. Often this involves noting that the function and/or its argument must be non-negative. For instance, suppose a word problem says that a builder wishes to use 100 feet of fence to enclose three sides of a rectangle, and asks for the rectangle's length as a function of its width. Such a function could be written as $f(x) = 100 - 2x$, where x is the width and $f(x)$ the length. Clearly, the width cannot be negative, so we must have $x \geq 0$. Nor can the length be negative, so we must have $f(x) \geq 0$, i.e. $100 - 2x \geq 0$, so $x \leq 50$. Therefore, the domain is $0 \leq x \leq 50$.

DETERMINING THE DOMAIN OF A FUNCTION FROM ITS GRAPH

The *domain* of a function is the set of x values for which the function is defined. This can be seen from the graph by observing for what values of x the function is drawn. Essentially, if a vertical line drawn through the graph at a particular value of x intercepts the graph of the function, then the function is defined at that point. If a vertical line intercepts the function no matter where it is drawn, then the domain contains all real numbers.

If this is not the case, then special considerations apply at the endpoints of intervals over which the function is defined. A function is never defined at a vertical asymptote. If $f(x)$ is finite at the endpoint of the interval, however, then the endpoint is drawn either as a filled circle to indicate that it is included in the function, or a hollow circle if not. Of the two graphs below, for instance, the left has a domain of $[-2, \infty)$, and the right of $(-2, \infty)$.

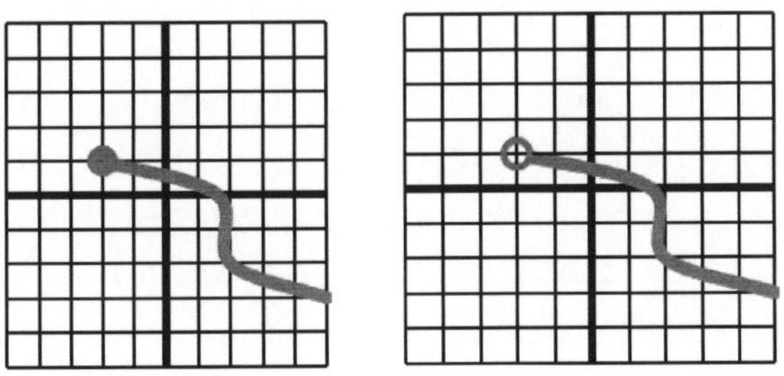

Example

Determine the domains of each of the functions shown below

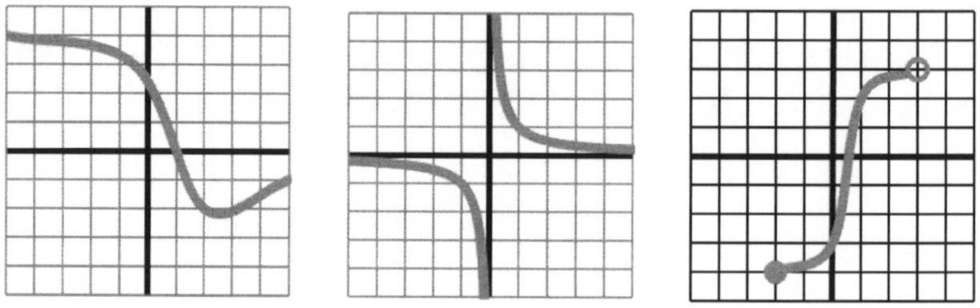

The domain of a function can be seen from its graph by observing what values along the x-axis have part of the function directly above or below them. Often the domain comprises all real numbers, $(-\infty, \infty)$. This is the case for the sample function shown on the left: the function extends unbroken from the left edge of the graph to its right edge; there is no x value for which it is not defined. For a function with a vertical asymptote, the x value corresponding to that asymptote is not part of the domain. The middle example function has a vertical asymptote at $x = 0$, so its domain is $(-\infty, 0) \cup (0, \infty)$.

Other functions are only defined over particular intervals. In this case, a solid circle at the end of the interval means that end point is included in the function's domain, and a hollow circle means it isn't. The example function on the right, then, is defined at $x = -2$ but not at $x = 3$; its domain is $[-2,3)$.

Example

Calculating a function's average rate of change over a specific interval when the function is given as a table

x	0	2	4	6	8	10	12	14	16
$f(x)$	-100	-54	-26	-10	0	10	26	54	100

The function's average rate of change over the interval $[x_1, x_2]$ can be calculated as $\frac{f(x_2)-f(x_1)}{x_2-x_1}$. For the sample table, for instance, if we want to determine the function's rate of change during the entire interval represented in the table, then $x_1 = 0$ and $x_2 = 16$. By reference to the table, $f(x_1) = f(0) = -100$, and $f(x_2) = f(16) = 100$. Thus, the function's rate of change is equal to $\frac{100-(-100)}{16-0} = \frac{200}{16} = 12.5$. If on the other hand we only want to find the rate of change between 0 and 4, then $x_1 = 0, x_2 = 4$, $f(x_1) = f(0) = -100$, and $f(x_2) = f(4) = -26$. The rate of change is then $\frac{-26-(-100)}{4-0} = \frac{74}{4} = 18.5$. If we are asked for the rate of change in the interval [6, 10], then $x_1 = 6, x_2 = 10, f(x_1) = f(6) = -10$, and $f(x_2) = f(10) = 10$. The rate of change is then $\frac{10-(-10)}{10-6} = \frac{20}{4} = 5$.

Example

Calculating a function's average rate of change over a specific interval when the function is given algebraically, such as $f(x) = x^2 + 1$

The function's average rate of change over the interval [x_1, x_2] can be calculated as $\frac{f(x_2)-f(x_1)}{x_2-x_1}$. For the sample function, for instance, suppose we're asked to find the function's average rate of change between 0 and 10. Then $x_1 = 0$ and $x_2 = 10$. To find $f(x_1)$ and $f(x_2)$, we simply evaluate the function at these two points: $f(x_1) = f(0) = 0^2 + 1 = 1$, and $f(x_2) = f(10) = 10^2 + 1 = 101$. The function's average rate of change in this interval is therefore $\frac{101-1}{10-0} = \frac{100}{10} = 10$.

If we're asked to find the function's average rate of change over the interval [-4, 2], then $x_1 = -4$, $x_2 = 2$, $f(x_1) = f(-4) = (-4)^2 + 1 = 17$, and $f(x_2) = f(2) = 2^2 + 1 = 5$. The average rate of change is therefore $\frac{5-17}{2-(-4)} = \frac{-12}{6} = -2$. If we're asked to find the average rate of change over the interval [-5, 5], then $x_1 = -5$, $x_2 = 5$, $f(x_1) = f(-5) = (-5)^2 + 1 = 26$, and $f(x_2) = f(5) = 5^2 + 1 = 26$. The average rate of change is therefore $\frac{26-26}{5-(-5)} = \frac{0}{10} = 0$.

ESTIMATING A FUNCTION'S AVERAGE RATE OF CHANGE OVER A SPECIFIC INTERVAL WHEN THE FUNCTION IS GIVEN GRAPHICALLY, LIKE THE GRAPH BELOW

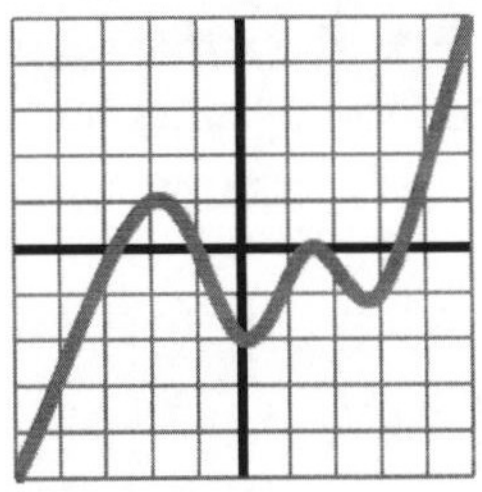

A function's average rate of change over an interval $[a, b]$ contained in the function's domain is the change in the value of $f(x)$ in that interval relative to the change in x. In other words, mathematically, it can be expressed as $\frac{f(b)-f(a)}{b-a}$. To estimate the average rate of change over an interval for a function given graphically, then, we have to estimate the coordinates of the endpoints of the desired interval and then put the appropriate values into that expression and carry out the calculation. For example, consider the sample function over the entire interval shown on the graph, $[-5,5]$. At the left end is passes through the point $(-5,-5)$, and at the right end $(5,5)$; the average rate of change is therefore $\frac{f(5)-f(-5)}{5-(-5)} = \frac{5-(-5)}{10} = \frac{10}{10} = 1$. If we instead consider the interval $[-2,0]$, the left endpoint is about $(-2,1)$ and the right about $(0,-2)$; the average rate of change over the interval $[-2, 0]$ is then $\frac{f(0)-f(-2)}{0-(-2)} = \frac{-2-1}{2} = \frac{-3}{2} = -\frac{3}{2}$. The average rate of change over the interval $[-5, 0]$ is $\frac{f(0)-f(-5)}{0-(-5)} = \frac{-2-(-5)}{5} = \frac{3}{5}$.

GRAPHING A LINEAR FUNCTION GIVEN IN SLOPE-INTERCEPT FORM, SUCH AS $f(x) = 3x + 1$

There are several ways to graph a linear function. For a function given in slope-intercept form, like this example, it's usually easiest to use the slope and the y intercept. The slope-intercept form of a linear function takes the form $y = mx + b$, where m is the slope and b the y intercept. (If the x has no coefficient, the slope is 1.) For the sample function $f(x) = 3x + 1$ (i.e. $y = 3x + 1$), the slope is 3 and the y-intercept is 1. Therefore, we can plot the point (0,1), and then draw a line through that point with the appropriate slope: for every 1 unit to the right, the line goes 3 units up. If the slope is negative, then instead of up and to the right, it goes down and to the right (or, equivalently, up and to the left). A graph of the sample function therefore looks as follows:

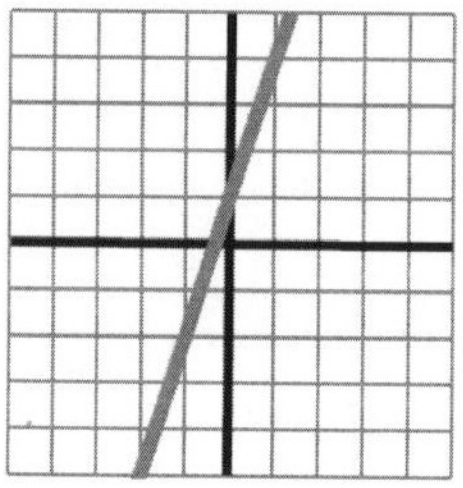

GRAPHING A LINEAR FUNCTION GIVEN IN STANDARD FORM, SUCH AS $2x - 3y = 12$

There are several ways to graph a linear function. If the function is given in standard form, $Ax + By = C$, like this example, then it's usually easiest to use the x and y intercepts. Consider the example equation, $2x - 3y = 12$. When $x = 0$, we get $-3y = 12$, so $y = -4$; this is the y intercept. When $y = 0$, we get $2x = 12$, so $x = 6$; this is the x intercept. We now know the points $(0, -4)$ and $(6, 0)$ are on the graph, so we can plot these two points and draw a straight line through them. A graph of the sample function therefore looks as follows:

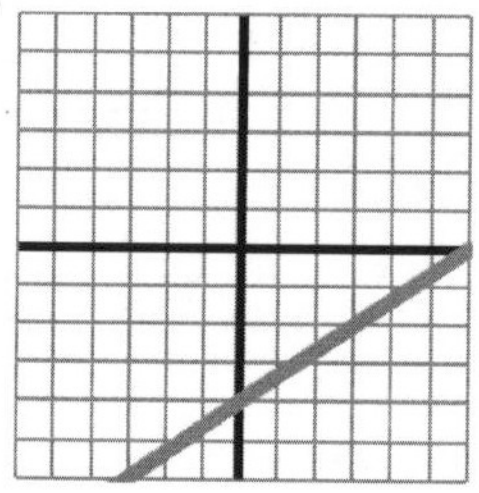

GRAPHING A QUADRATIC FUNCTION, SUCH AS $f(x) = 2(x - 4)^2 + 3$

If a quadratic function is in *vertex form*, $a(x - h)^2 + k$, we can see at a glance the coordinates of the vertex: they are simply (h, k). In this case, $h = 4$ and $k = 3$, so the vertex is at (4,3). We can use the coefficient a to see the direction of the opening: if $a > 0$, the graph opens upward, while if $a < 0$ it opens downward. Here $a = 2 > 0$, so this graph opens upward. After that, the only remaining feature to distinguish the quadratic function is its width. We can get a feel for this by simply plotting a few other points. The x intercepts are good choices, but some quadratic functions have no x intercepts, like this one: the equation $0 = 2(x - 4)^2 + 3$ has no real solutions. However, we can still choose other values to plot: $f(3) = 2(3-4)^2 + 3 = 5$ and $f(5) = 2(5-4)^2 + 3 = 5$, so we can plot the points (3,5) and (5,5). Our final graph in this case looks like this:

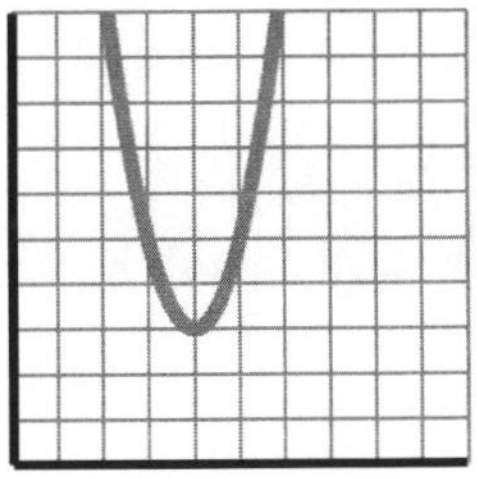

PIECEWISE-DEFINED FUNCTION

A *piecewise-defined function* is a function that has different definitions on two or more different intervals. The following, for instance, is one example of a piecewise-defined function:

$$f(x) = \begin{cases} x^2, & x < 0 \\ x, & 0 \le x \le 2 \\ (x-2)^2, & x > 2 \end{cases}$$

To graph this function, we'd simply graph each part separately in the appropriate domain. Our final graph would look like this:

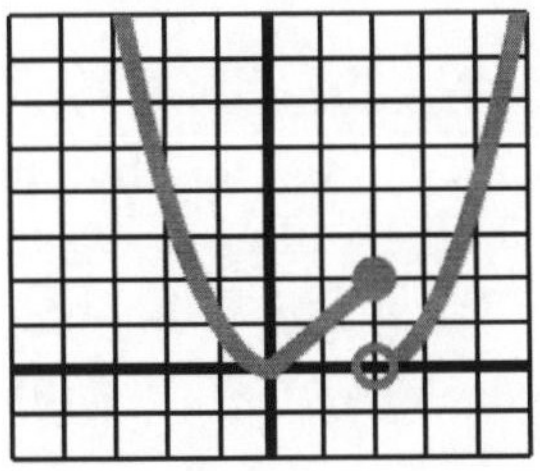

Note the filled and hollow dots at the discontinuity at $x = 2$. This is important to show which side of the graph that point corresponds to. Because $f(x) = x$ on the closed interval $0 \le x \le 2$, $f(2) = 2$. The point (2,2) is therefore marked with a filled circle, and the point (2,0), which is the endpoint of the rightmost $(x-2)^2$ part of the graph but *not actually part of the function*, is marked with a hollow dot to indicate this.

GRAPHING A SQUARE ROOT FUNCTION, SUCH AS $f(x) = -2\sqrt{4x + 4}$

All square root functions have a characteristic shape (essentially, the shape of half a quadratic function rotated ninety degrees). They vary, however, in their dimensions and displacement. If we rewrite the given function as $f(x) = -2\sqrt{4(x + 1)}$, we can see by inspection how it differs from the "plain" square root function $f(x) = \sqrt{x}$: it is shifted one unit to the left (because of the $x + 1$), flipped over vertically (because of the sign of the coefficient -2), stretched vertically by a factor of 2, and squashed horizontally by a factor of 4. Alternately, we can just plot a few points to get a feel for the graph's shape: for instance, $f(-1) = -2\sqrt{-4 + 4} = 0$, $f(0) = -2\sqrt{0 + 4} = -4$, and $f(1) = -2\sqrt{4 + 4} = -4\sqrt{2} \approx -5.66$. Our final graph looks like this:

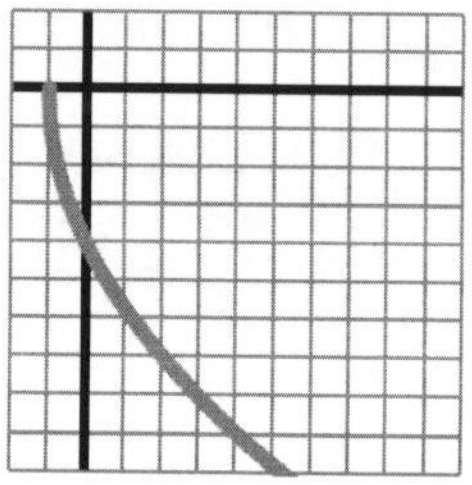

GRAPHING A CUBE ROOT FUNCTION SUCH AS $f(x) = 3\sqrt[3]{x} - 2$

All cube root functions have a characteristic shape (somewhat like a stretched-out S—note that unlike that of the square root, the domain of the cube root function encompasses all real numbers). They vary, however, in their dimensions and displacement. In the case of the given function, $3\sqrt[3]{x} - 2$, we can see by inspection how it differs from the "plain" cube root function $f(x) = \sqrt[3]{x}$: it is stretched vertically by a factor of 3 (because of the coefficient 3), and shifted two units downward (because of the -2). We can just draw a cube root function modified accordingly. Alternately, we can just plot a few points to get a feel for the graph's shape: for instance, $f(0) = 3\sqrt[3]{0} - 2 = -2$, $f(1) = 3\sqrt[3]{1} - 2 = 1$, $f(8) = 3\sqrt[3]{8} - 2 = 4$, and so on. Our final graph looks like this:

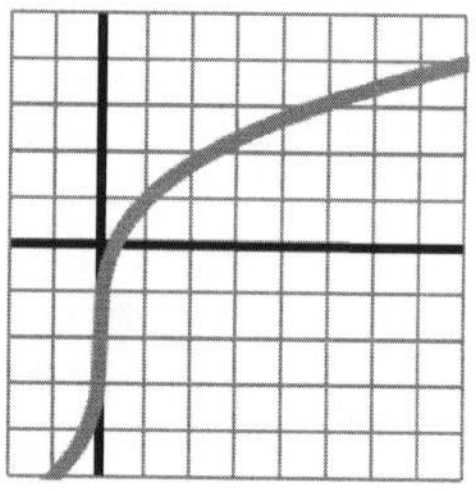

GRAPHING AN ABSOLUTE VALUE FUNCTION SUCH AS $f(x) = 2|x + 1| - 3$

The key to graphing an absolute value function is to recognize that its behavior depends on whether the expression within the absolute value sign is positive or negative, and to graph these parts separately. In the case of the function $f(x) = 2|x + 1| - 3$, the point at which the behavior changes is when $x + 1 = 0$, i.e. when $x = -1$. When $x \geq -1$, $|x + 1| = x + 1$, and the function reduces to the linear function $f_+(x) = 2(x + 1) - 3 = 2x - 1$. When $x \leq -1$, $|x + 1| = -(x + 1)$, and the function reduces to $f_{-}(x) = 2(-(x+1)) - 3 = -2x - 5$. So we can graph $y = -2x - 5$ to the left of $x = -1$ and $y = 2x - 1$ to the right, yielding the following graph:

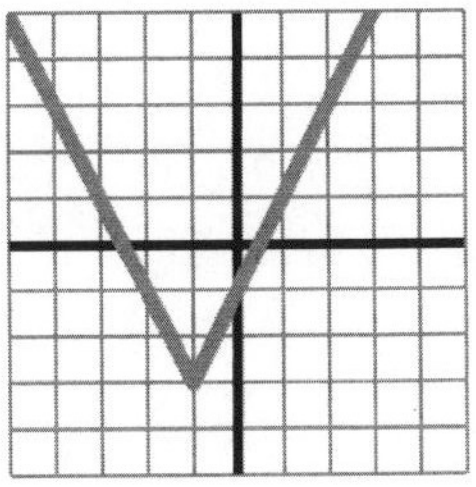

GRAPHING A STEP FUNCTION SUCH AS $f(x) = 2\left[\!\left[\frac{1}{3}(x-1)\right]\!\right]$

The double brackets indicate a step function. For a step function, all values inside the double brackets are rounded down to the nearest integer. The graph of the function $f_0(x) = [\![x]\!]$ appears as shown on the graph on the left below. In this case, we can see by inspection how the given $f(x) = 2\left[\!\left[\frac{1}{3}(x-1)\right]\!\right]$ differs from $f_0(x)$. The coefficient of 2 shows that it's stretched vertically by a factor of 2 (so there's a vertical distance of 2 units between successive "steps"). The coefficient of $\frac{1}{3}$ in front of the x shows that it's stretched horizontally by a factor of 3 (so each "step" is three units long), and the $x - 1$ shows that it's displaced one unit to the right. The final graph, then, is on the right below.

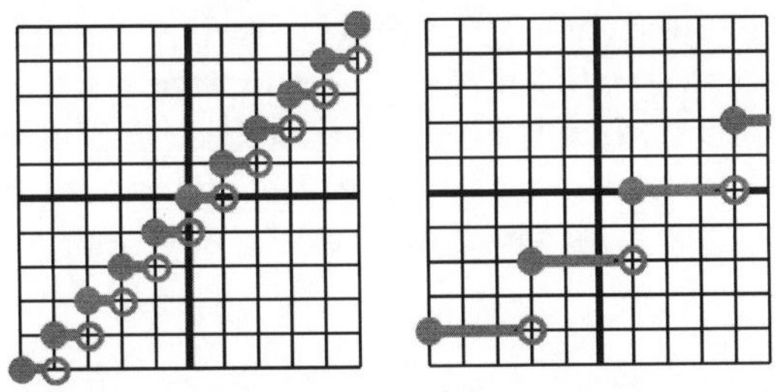

DETERMINING THE END BEHAVIOR OF A POLYNOMIAL FUNCTION

The "end behavior" of a function is what happens to the function when $|x|$ is very large. For a polynomial function, there are only two possibilities: either the function continues increasing indefinitely (it "goes to infinity"), or the function continues *de*creasing indefinitely (it "goes to negative infinity").

To see which is true, we need only look at the largest-order term of the polynomial, that is, the term with the largest exponent. To find the behavior of the function as $x \to \infty$ (at the right end), we just look at the coefficient of that term: if it's positive, the function goes to infinity; if it's negative, the function goes to negative infinity. The behavior of the function as $x \to -\infty$ (at the left end) is only slightly more complicated. If the exponent is even, the same rule holds as for the right end. If the exponent is odd, they're reversed: if the coefficient is positive, the function goes to negative infinity, and if it's negative, the function goes to positive infinity.

GRAPHING A POLYNOMIAL SUCH AS $\frac{1}{4}x^4 - \frac{9}{4}x^2$

To graph a polynomial, like most other functions, there are a few key characteristics we can observe. First, we can observe the end behavior, which depends only on the largest order term. To find the y intercept, we evaluate the function at $x = 0$. To find the x intercepts, the points where the polynomial equals zero, is a little harder, but often we can factor the function and set each factor equal to zero. Depending on the polynomial, there may be other characteristics we can find, such as the local maxima and minima.

For the given example, since the highest order term has an even order and a positive coefficient, the polynomial goes to positive infinity at both ends. The *y* intercept is at $\frac{1}{4}(0^4) - \frac{9}{4}(0^2) = 0$. We can factor the polynomial as $\frac{1}{4}x^4 - \frac{9}{4}x^2 = \frac{1}{4}x^2(x^2 - 9) = \frac{1}{4}x^2(x + 3)(x - 3)$, so it has *x* intercepts at $x = 0$, 3, and –3. Its graph looks like this:

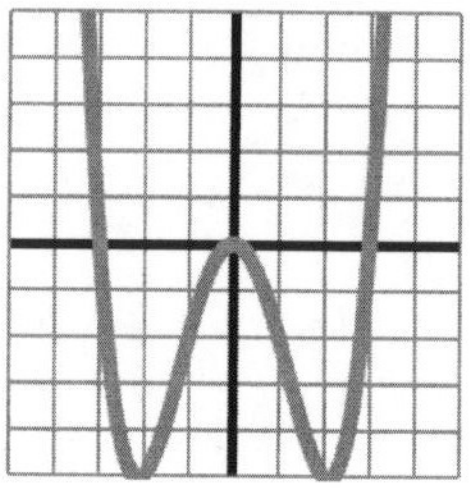

Example

Identifying the zeroes of a polynomial such as $x^3 - 3x^2 - 4x$ or $x^3 + 2x^2 - 2x - 4$

We can find the zeroes of a polynomial by factoring it. Cubics and higher polynomials can be difficult to factor, but there are special cases that can be readily factored. In the first sample polynomial here, for instance, $x^3 - 3x^2 - 4x$, it's clear that every term of the polynomial is divisible by *x*. We can therefore factor out the *x* to get $x(x^2 - 3x - 4)$. This now leaves a quadratic factor, which can be further factored to yield $x(x + 1)(x - 4)$. The zeroes of the polynomial are therefore 0, –1, and 4.

In the second polynomial, $x^3 + 2x^2 - 2x - 4$, there's no such divisor that goes into every term. If we compare the first two terms with the last two, however, we note that both *pairs* of terms are divisible by $x + 2$. We can then write the polynomial as $x^2(x + 2) - 2(x + 2)$, or $(x^2 - 2)(x + 2)$, and further factoring the first term,

$(x + \sqrt{2})(x - \sqrt{2})(x + 2)$. The zeroes of this polynomial are –2 and $\pm\sqrt{2}$.

Finding the Horizontal Asymptote of a Rational Function

A rational function is a ratio of polynomials, and the horizontal asymptote is a horizontal line that the rational function tends toward as *x* approaches $\pm\infty$. To find the horizontal asymptote, first compare the orders of the numerator and the denominator—that is, the largest exponents appearing in them. If the numerator has a larger order than the denominator (as in $\frac{x^2-1}{x}$ or $\frac{x^4+x+2}{x^2-x}$), then the function has no horizontal asymptote. If the numerator has a smaller order the denominator (as in $\frac{x}{x^2-1}$ or $\frac{x^2+x+1}{x^5}$), then the function has a horizontal asymptote at $y = 0$.

If the orders of the numerator and denominator are equal (as in $\frac{2x^2+2}{x^2+2x+1}$ or $\frac{3x-4}{4x-3}$), then the horizontal asymptote is $y = y_0$, where y_0 is the ratio of the coefficients of the highest-order terms.

For instance, the rational function $\frac{2x^2+2}{x^2+2x+1}$ has a horizontal asymptote at $y = \frac{2}{1} = 2$; $\frac{3x-4}{4x-3}$ has a horizontal asymptote at $y = \frac{3}{4}$.

Determining Whether a Rational Function Has a Horizontal Asymptote, a Slant Asymptote, or Neither

To determine the end behavior of a rational function, it is only necessary to compare the orders of the numerator and the denominator—that is, the largest exponents appearing in them. If the order of the numerator is less than or equal to the order of the denominator, then the function has a horizontal asymptote. If the order of the numerator is one more than the order of the denominator, then the function has a slant asymptote. If the order of the numerator exceeds the order of the denominator by two or more, then the function has neither a horizontal nor a slant asymptote.

For instance, consider the rational function $\frac{3x^3+4x^2-2}{5x^7-5}$. The order of the numerator is 3; the order of the denominator is 7. Since $3 \leq 7$, this function has a horizontal asymptote (at $y = 0$). In the rational function $\frac{2x^4-x^2+3}{x^3-x}$, the order of the numerator is 4 and the order of the denominator is 3. Since $4 = 3 + 1$, this function has a slant asymptote.

Finding the Vertical Asymptotes of a Rational Function

Generally, a rational function has a vertical asymptote where the denominator is equal to zero, making the function undefined at that point (since the result of dividing by zero is undefined). For instance, consider the function $\frac{x^2+4x+4}{x^3-1}$. $1^3 - 1 = 0$, so this function has a vertical asymptote at $x = 1$.

There is one important exception to the above rule, however. If there is a common term to the numerator and denominator of the rational function, then it has vertical asymptotes only where the denominator is zero *after canceling common terms*. For instance, the function $f(x) = \frac{x^2-2x+1}{x^2+x-2}$ is undefined at $x = 1$, since

$1^2 + 1 - 2 = 0$. However, $\frac{x^2-2x+1}{x^2+x-2} = \frac{(x-1)(x-1)}{(x-1)(x+2)}$; canceling the common term yields $\frac{x-1}{x+2}$. So $f(x)$ has a vertical asymptote at $x = -2$, but *not* at $x = 1$. (This doesn't change the fact that $f(x)$ is *undefined* at $x = 1$, but it can be undefined there without having a vertical asymptote.)

Determining the end behavior of a rational function such as $f(x) = \frac{3x^2+2x-3}{6x^3-2x^2+1}$ or $g(x) = \frac{-7x^7+1}{2x^2-x+3}$

To determine the end behavior of a rational function, it's only necessary to consider the leading terms of the numerator and denominator—the terms with the largest exponents. If the leading term of the denominator has a larger exponent, then $f(x)$ goes to zero as x goes to $\pm\infty$. We can then take the ratio of these leading terms, which will result in a single term of which the end behavior should be clear. For example, consider the first example, $f(x) = \frac{3x^2+2x-3}{6x^3-2x^2+1}$. The leading term of the numerator is $3x^2$, and the leading term of the denominator is $6x^3$. The latter has a larger exponent, so this function goes to zero as x goes to $\pm\infty$. In the second function, $g(x) = \frac{-7x^7+1}{2x^2-x+3}$, the leading term of the numerator is $-7x^7$, and the leading term of the numerator is $2x^2$. The ratio is $\frac{-7x^7}{2x^2} = -\frac{7}{2}x^5$, which goes to ∞ as x goes to $-\infty$ and $-\infty$ as x goes to ∞.

Finding the Zeroes of a Rational Function such as $f(x) = \frac{x^2+3x+2}{x^3+x+1}$ or $g(x) = \frac{x^4-x^2}{x^3-2x^2+1}$.

A rational function $R(x)$ is a ratio of two polynomials, which we can write as $\frac{P(x)}{Q(x)}$. The rational function has a zero at any point $x = x_0$ for which the numerator $P(x_0)$ is equal to zero and the denominator $Q(x_0)$ is *not* equal to zero. If $Q(x_0) = 0$, then $R(x_0)$ is undefined.

In the first sample function, $f(x) = \frac{x^2+3x+2}{x^3+x+1}$, the numerator $P(x) = x^2 + 3x + 2$ has zeroes at $x = -1$ and $x = -2$. Neither of these is a zero of the denominator $Q(x) = x^3 + x + 1$: $Q(-1) = -1$, and $Q(-2) = -9$. So both of these are zeroes of $f(x)$.

In the second sample function, $g(x) = \frac{x^4-x^2}{x^3-2x^2+1}$, the numerator $P(x) = x^4 - x^2 = x^2(x^2 - 1)$ has zeroes at $x = 0$ and $x = \pm 1$. Before concluding that these are zeroes of $g(x)$, though, we have to check whether they're zeroes of the denominator $Q(x) = x^3 - 2x^2 + 1$. $Q(-1) = -2$, $Q(0) = 1$, and $Q(1) = 0$. This means $x = 1$ is *not* a zero of $g(x)$, so the zeroes of$g(x)$ are $x = -1$ and $x = 0$.

Example

> Given a trigonometric function of the form $f(x) = A\sin(Bx + C) + D$ (such as $f(x) = 2 + \frac{3}{2}\sin\left(\pi x + \frac{\pi}{2}\right)$), determine how you would find the function's period, amplitude, and midline and use them to graph the function
>
> In the function $f(x) = A\sin(Bx + C) + D$, the amplitude is simply equal to A, and the midline is $y = D$. The period is only a little more complicated; the period is equal to ${2\pi}/{B}$. For instance, consider the example function $f(x) = 2 + \frac{3}{2}\sin\left(\pi x + \frac{\pi}{2}\right)$. Here $A = \frac{3}{2}$, $B = \pi$, $C = \frac{\pi}{2}$, and $D = 2$, so the midline is at $y = 2$, the amplitude is $\frac{3}{2}$, and the period is ${2\pi}/{\pi} = 2$.
>
> To graph this equation, we center the sine wave on the midline and extend it to a height above and below the midline equal to the amplitude—so this graph would have a minimum value of $2 - \frac{3}{2} = \frac{1}{2}$ and a maximum of $2 + \frac{3}{2} = \frac{7}{2}$. The period (here equal to 2) is the distance between successive peaks or troughs. As for the last value, C, this is related to the *phase shift*, which is equal to $-\frac{C}{B}$ (in this case $-\frac{1}{2}$) and can be

thought of as a starting point where $f(x) = D$ (the midline) and increasing. So the function above would be graphed as follows:

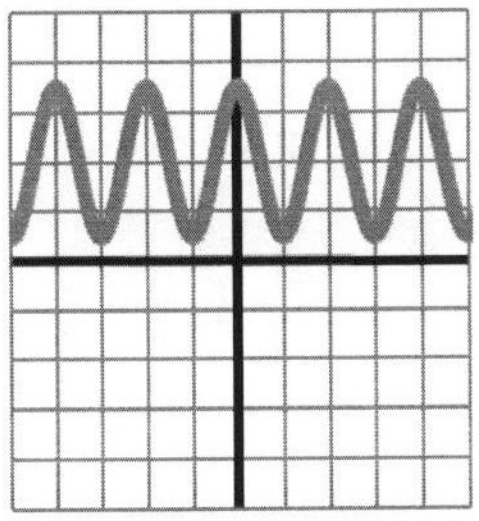

Example

Find the x and y intercepts, if any, of an exponential function of the form $f(x) = Ae^{bx} + C$

To find the y intercept, we just evaluate the function at $x = 0$. This gives $f(0) = Ae^{b\cdot 0} + C = Ae^0 + C = A(1) + C = A + C$. For instance, if $f(x) = \frac{1}{2}e^{3x} - 4$, $f(0) = \frac{1}{2}e^{3\cdot 0} - 4 = \frac{1}{2}(1) - 4 = -\frac{7}{2}$. This function has a y intercept at $(0, -\frac{7}{2})$.

To find the x intercept, we have to find a value of x for which $f(x) = 0$. Thus, $Ae^{bx} + C = 0$, so $Ae^{bx} = -C$, and $e^{bx} = -\frac{C}{A}$. From here, we have to take the logarithm of both sides, so $bx = \ln\left(-\frac{C}{A}\right)$, and $x = \frac{1}{b}\ln\left(-\frac{C}{A}\right)$. (Note that the logarithm of a negative number is undefined, so this has a solution only if $-\frac{C}{A} > 0$, i.e. $\frac{C}{A} < 0$. Otherwise, there is no x intercept.) If $f(x) = \frac{1}{2}e^{3x} - 4$, then to find the x intercept we use $\frac{1}{2}e^{3x} - 4 = 0$, so $\frac{1}{2}e^{3x} = 4$, $e^{3x} = 8$, and hence $3x = \ln 8$ and $x = \frac{\ln 8}{3} \approx 0.693$.

Example

Determine the end behavior of an exponential function such as $f(x) = 2e^{2x}$ or $g(x) = -e^{-x}$

The exponential function $y = e^x$ goes to zero as x goes to $-\infty$ and goes to ∞ as x goes to ∞. Multiplying either the exponent or the full exponential by a positive coefficient doesn't change this behavior, so this is the end behavior of $f(x) = 2e^{2x}$: it goes to zero on the left and goes to ∞ on the right.

Negative coefficients, on the other hand, do make a difference. A negative coefficient in the exponent interchanges the left and right side behavior, while a negative coefficient in front of the entire function makes the function go to negative infinity on the right instead of positive infinity. The function $g(x) = -e^{-x}$ has negative coefficients in *both* places, so it goes to $-\infty$ as x goes to $-\infty$, and goes to zero as x goes to ∞.

Example

Determine the end behavior of a logarithmic function such as $f(x) = \ln(-x)$ or $g(x) = -\frac{1}{2}\log_{10}(x+2)$

Regardless of the base, the logarithmic function $y = \log(x)$ goes to ∞ as x goes to ∞. The domain of $y = \log(x)$ is $(0, \infty)$, so we don't need to consider what happens as x goes to $-\infty$, but as x approaches the vertical asymptote at $x = 0$, $\log(x)$ goes to $-\infty$.

The function $f(x) = \ln(-x)$ has a negative coefficient in front of the x, which means the function is reflected from left to right. This function, therefore, goes to ∞ as x goes to $-\infty$, and on the right $f(x)$ goes to $-\infty$ as x approaches the vertical asymptote at $x = 0$.

The $x + 2$ in the function $g(x) = -\frac{1}{2}\log_{10}(x+2)$ means the function is shifted two units to the left—which means, in particular, that the vertical asymptote is shifted two units to the left. The negative coefficient in front of the function means the entire graph is flipped vertically. Therefore, $g(x)$ goes to $-\infty$ as x goes to ∞, and goes to ∞ as x approaches the vertical asymptote at $x = -2$.

Example

Given a tangent function such as $f(x) = 2\tan(3x - 4) + 1$, determine how you would find the function's period and midline and use them to graph the function

In the function $f(x) = A\tan(Bx + C) + D$, the midline is $y = D$, and the period is $^{\pi}/_{B}$. Larger values of A stretch the tangent vertically, but since the range of the tangent includes all real numbers, the tangent doesn't have an amplitude in the same sense as the sine and cosine do. If A is negative, the function is flipped vertically. C is related to the *phase shift*, ϕ, which is equal to $-C/B$, and is the amount by which the function is displaced horizontally.

For instance, consider the example function $f(x) = 2\tan(3x - 4) + 1$. Here $A = 2$, $B = 3$, $C = -4$, and $D = 1$, so the midline is at $y = 1$, the period is $^{\pi}/_{3} \approx 1.05$, and the phase shift is $-(-{}^{4}/_{3}) = {}^{4}/_{3}$. To graph this equation, we start at the point (ϕ, D)—in this case, $({}^{4}/_{3}, 1)$—and we draw the tangent shape centered on this point, with a width equal to the period. We then repeat this shape to both sides at regular intervals. The result is as follows:

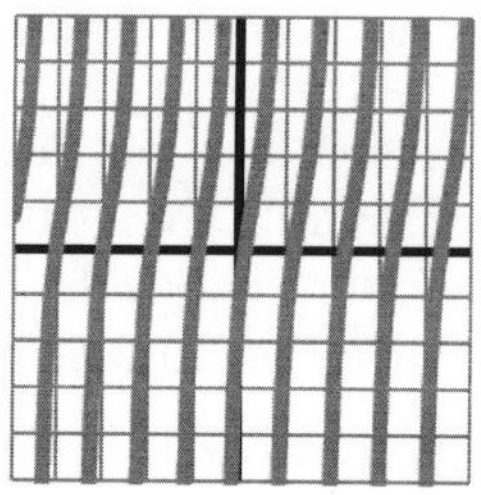

What it Means to "Complete the Square" of a Quadratic Function, and What Purpose it Serves

A quadratic function in the "vertex form" $f(x) = a(x-h)^2 + k$ readily shows the coordinates of the vertex, (h, k). However, quadratic functions are often given in "standard form", $f(x) = ax^2 + bx + c$, which doesn't show the coordinates of the vertex directly. Completing the square is one way to convert the function from standard form to vertex form. To complete the square, we first factor out the a from the first two terms: $f(x) = a\left(x^2 + \frac{b}{a}x\right) + c$. Now, we have to find a linear expression which, squared, yields a quadratic the first two terms of which match the terms in parentheses. This is probably most easily shown with an example: if $f(x) = 2x^2 + 8x + 6 = 2(x^2 + 4x) + 6$, then a suitable linear expression is $x + 2$, because $(x+2)^2 = x^2 + 4x + 4$. Therefore, we have $f(x) = 2(x+2)^2 + h$. All we have to do now is find a value for h that matches the original equation. Expanding, we have

$f(x) = 2x^2 + 8x + 8 + h$, so we require $8 + h = 6$, thus $h = -2$, and $f(x) = 2(x+2)^2 - 2$.

Factoring a quadratic function such as $f(x) = x^2 + x - 6$, $g(x) = x^2 - 16$, or $h(x) = 2x^2 + 3x - 7$

Factoring a quadratic function of the form $ax^2 + bx + c$ means finding two factors $(px + q)$ and $(rx + s)$ such that $(px+q)(rx+s) = ax^2 + bx + c$. Since $(px+q)(rx+s) = prx^2 + (ps+qr)x + qs$, we're looking specifically for p, q, r, s such that $pr = a$, $ps + qr = b$, and $qs = c$. The matter is much simpler if $a = 1$; then we need only find q and s such that $q + s = b$ and $qs = c$. For instance, for the first example, $f(x) = x^2 + x - 6$, $b = 1$ and $c = -6$, so we need to find two numbers that add to 1 and multiply to -6. -2 and 3 qualify, so $f(x)$ can be factored to $(x-2)(x+3)$. When the quadratic function has the form $x^2 - c$, we don't even need to do that much work; we know the function's factors are $(x + \sqrt{c})$ and $(x - \sqrt{c})$. So $g(x) = x^2 - 16$ factorizes to $(x + \sqrt{16})(x - \sqrt{16}) = (x+4)(x-4)$.

Sometimes a quadratic function can't readily be factored by any of these systems. In this case, one can resort to the *quadratic formula*: $x = \frac{-b \pm \sqrt{b^2 - 4ac}}{2a}$. Note that when you use the quadratic formula to solve the equation $0 = ax^2 + bx + c$ what you are really solving is $0 = x^2 + \frac{b}{a}x + \frac{c}{a}$; therefore, when using the quadratic formula to find the factors of $ax^2 + bx + c$ you must include the coefficient a as one of your factors. In the case of our third sample function, $h(x) = 2x^2 + 3x - 7$, this gives us $x = \frac{-3 \pm \sqrt{3^2 - 4(2)(-7)}}{2 \cdot 2} = \frac{-3 \pm \sqrt{65}}{4} = -\frac{3}{4} \pm \frac{\sqrt{65}}{4}$, and the factorization is $2\left(x + \frac{3}{4} - \frac{\sqrt{65}}{4}\right)\left(x + \frac{3}{4} + \frac{\sqrt{65}}{4}\right)$.

Finding the Extreme Values of a Quadratic Function such as $f(x) = x^2 - 6x + 3$

A quadratic function of the form $ax^2 + bx + c$ has only a single extreme value, at its vertex. Whether this is a maximum or a minimum depends on whether the graph opens upward or downward. This in turn depends on the sign of the coefficient a: if a is positive, then the graph opens upward and the vertex is a minimum, while if a is negative, the graph opens downward and the vertex is a maximum.

In the case of the sample function $f(x) = x^2 - 6x + 3$, the coefficient $a = 1 > 0$, so the vertex is a minimum. To find the coordinates of the vertex, we can put the function in *vertex form*, $f(x) = a(x-h)^2 + k$. We can do this by completing the square: $(x-3)^2 = x^2 - 6x + 9$, so $f(x) = (x-3)^2 - 6$. The vertex is at the coordinates (h, k)—in this case, $(3, -6)$.

SYMMETRY OF A QUADRATIC FUNCTION SUCH AS $f(x) = 3(x - 2)^2 + 5$

Any quadratic function is symmetrical through a vertical line passing through the vertex. When the quadratic function is in vertex form, $f(x) = a(x - h)^2 + k$, the coordinates of the vertex can be readily seen from the equation; they are simply (h, k) . The axis of symmetry is then $x = h$; the function is unchanged when reflected through this line. If $h = 0$, then the axis of symmetry is the line $x = 0$, which is the y axis; the function is then symmetrical with respect to reflections through the y axis, which means it is an even function. Quadratic functions with $h \neq 0$ are neither even nor odd.

In the case of the sample function given here, $f(x) = 3(x - 2)^2 + 5, h = 2$ and $k = 5$. This function, therefore, is symmetrical with respect to reflections through the line $x = 2$.

Example

Given an exponential equation such as $f(x) = 200 \cdot (1.2)^{3x}$, show how you can determine whether it represents exponential growth or exponential decay

If the coefficient of the exponent is positive (as it is in the sample function given), then the function represents exponential growth if the base of the exponent is greater than 1, and exponential decay if the exponent is less than 1. The sample function therefore represents exponential growth, since the base, 1.2, is greater than 1. An example of a function that would represent exponential decay would be $g(x) = 13 \cdot (0.9)^{2.5x}$, since here the base, 0.9, is less than one.

If the coefficient in the exponent is negative, then this rule of thumb is reversed. Thus, for instance,

$h(x) = 10 \cdot (1.1)^{-2x}$ represents exponential decay, since the coefficient (-2) is negative, and the base (1.1) is greater than 1; $k(x) = 7 \cdot (0.8)^{-1.5x}$ is exponential growth, since the exponent (-1.5) is negative, and the base (0.8) is less than 1.

Example

Finding the percent rate of change of an exponential function such as $f(t) = 70 \cdot (1.4)^{1.05t}$

The percent rate of change is the amount by which a function changes per unit of time, expressed as a percentage. This can be calculated as $\frac{f(t+1)-f(t)}{f(t)}$. For the sample function, $f(t) = 70 \cdot (1.4)^{1.05t}$, this calculation gives $\frac{70 \cdot (1.4)^{1.05(t+1)} - 70 \cdot (1.4)^{1.05t}}{70 \cdot (1.4)^{1.05t}} = \frac{70 \cdot (1.4)^{1.05(t+1)}}{70 \cdot (1.4)^{1.05t}} - 1 = (1.4)^{(1.05(t+1) - 1.05t)} - 1 = 1.4^{1.05} - 1 \approx 0.424 = 42.4\%$ per unit of time. Note that the time, t, cancels from the calculation, and therefore the rate of change does not depend on t. This is a characteristic of exponential functions; they have a constant percent rate of change.

The answer we obtained for the sample function can be generalized: in general, the percent rate of change of an exponential function $f(t) = a \cdot b^{ct}$ is equal to $b^c - 1$. Note that if $b^c < 1$, then the percent rate of change is negative. For instance, consider $g(t) = 2 \cdot \left(\frac{1}{2}\right)^t$. The percent rate of change is $\left(\frac{1}{2}\right)^1 - 1 = -\frac{1}{2} = -50\%$. This makes sense, because if $b^c < 1$ then the function represents exponential decay; the function is decreasing, so the percent rate of change is indeed negative.

Important Properties of Exponents, and Explain How They Could Be Used in a Problem

One important property of exponents is the property that $x^a x^b = x^{a+b}$. This could come into play in a problem in which we have a product of two exponentials, and it simplifies the problem to combine them into one. Conversely, we can use it to simplify an expression with a sum in the exponent, such as $3 \cdot 2^{x+4}$; this simplifies to $3 \cdot 2^x 2^4 = 3 \cdot 2^x \cdot 16 = 48 \cdot 2^x$.

Often given separately is the property $x^a/_{x^b} = x^{a-b}$, but this is really a special case of the previous property where $b < 0$, and applies under similar circumstances.

Another important property of exponents is the property that $(x^a)^b = x^{ab}$. This can be used to simplify expressions in which an exponential is raised to a power, or to simplify an expression with a product in the exponent into a single exponent: for example, $3^{3x} = (3^3)^x = 27^x$.

Given Two Functions, One Represented Algebraically and One Graphically, Determine How You Can Find Which Has the Larger Maximum

One way to compare the features of two functions represented in different ways is to convert them both to the same manner of representation. If one function is represented algebraically and the other graphically, we can graph the first function on the same axis as the second, enabling us to compare their maxima visually by just observing which one extends farther in the positive y direction.

It's also possible to compare their maxima without converting them to the same representation, by simply finding the maximum of each function and comparing them. If the function expressed algebraically is one for which we can find the maximum analytically (for example, a quadratic function), we can compare that value directly with the maximum estimated from the graphed function by observing the y coordinate of its highest peak.

Given Two Functions, One of Which Is Given Algebraically and the Other in a Verbal Description, Determine How You Could Find Which Has the Larger Maximum

When given two functions in different formats, one way to compare their properties is to convert both functions to the same format. If one of the functions is given in the form of a verbal description, we can write an algebraic expression to represent the described relationship. For instance, if we're told that a function represents the area of a rectangular field for which the total distance around three sides is 40 meters (implying that, if we write l as the length and w as the width, $l + 2w = 40$), we can write the relationship between the area and the width algebraically as $A = l \cdot w = (40 - 2w)w$. Or, using the more familiar f and x, $f(x) = (40 - 2x)x = -2x^2 + 40x$. Completing the square gives us $f(x) = -2(x - 10)^2 + 200$. This function has a maximum when $x = 10$, at which point $f(x) = 200$. We can likewise find the maximum of the other function given algebraically, and we can now compare the maxima directly.

Given Two Functions, One of Which Is Given Graphically and One in a Table, Determine How You Can Find Which Has the Larger Y Intercept

When given two functions in different formats, one way to compare their properties is to convert both functions to the same format. If one of the functions is given as a table, we can graph that function and then compare it to the graph directly. Conversely, we could determine the values of the graphed function at the x values given in the table and therefore express both functions in tabular format.

It's also sometimes possible to compare properties of the functions even without converting them to the same format, and in the case of finding the y intercept this isn't difficult (assuming that the y intercept of the second function is actually given in the table). We can find the y intercept of the graphed function by seeing where the graph crosses the y-axis, find the y intercept of the table function by seeing what value of the function corresponds to an x value of 0, and compare these values directly.

Example

> Given a description of a relationship, such as a statement that the population of a particular area is 2000 in the year 1980 and doubles every 10 years thereafter, determine how you can write an explicit expression to describe that relationship
>
> The first step in writing an expression to describe a relationship is to pinpoint the variables involved in the expression. In the sample case, the statement describes how the population changes over time, we will write the population P as a function of time, t.
>
> Next, we look for telltale signs that tell us what kind of function we are dealing with. Here, the population increases by a fixed proportion over a specific unit of time, a hallmark of an exponential relationship. Therefore, the expression will take the form $P(t) = P_0 A^{ct}$. More specifically, since we're given the population at 1980 as a starting point, we can simplify matters by setting the exponent proportional to the number of years since 1980: $P(t) = P_{1980} A^{c(t-1980)}$.
>
> Finally, we use the given data to put in numbers. When $t = 1980$, $P(t) = 2000$, so $P_{1980} = 2000$. Incorporating the other information gives us $P(t) = 2000 \cdot 2^{(t-1980)/10}$.

DETERMINE WHAT KINDS OF RELATIONSHIPS LEND THEMSELVES TO BEING WRITTEN AS RECURSIVELY DEFINED FUNCTIONS

A recursively defined function is generally a *sequence*; that is, its domain consists of a subset of the integers (usually, though not always, the positive integers). Furthermore, the value of the function for a given value of x must be defined in terms of its value for smaller values of x. Often this means $f(x)$ is defined in terms of $f(x-1)$, though it may also be defined in terms of even smaller values. Finally, we need to have $f(x_0)$ explicitly defined for the smallest value or values of x_0 to give us a starting point.

For example, consider the following relationship: A pile of stones starts with three stones, and then every day its size is doubled and then one stone is removed. On day one, then, the size of the pile is $f(1) = 3$. On day x thereafter, we have the relationship $f(x) = 2f(x-1) - 1$. This is a valid recursive definition.

Example

> Suppose $f(x) = x^2 - 1$ and $g(x) = x - 1$. How would you find each of the following
>
> $(f+g)(x)$ $\quad$ $(f-g)(x)$ $\quad$ $(f \cdot g)(x)$ $\quad$ $(f/g)(x)$

To combine functions using arithmetic operations, we need only apply the appropriate arithmetic operation to the function's expression. So $(f+g)(x) = f(x) + g(x) = (x^2 - 1) + (x - 1) = x^2 + x - 2$. $(f-g)(x) = (x^2 - 1) - (x - 1) = x^2 - x$. $(f \cdot g)(x) = (x^2 - 1)(x - 1) = x^3 - x^2 - x + 1$.

Finally, $(f/g)(x) = (x^2 - 1)/(x - 1)$. Here we have to be a little more careful. $(x^2 - 1) = (x + 1)(x - 1)$, so it's tempting to cancel out the $(x - 1)$ terms and state simply that $(f/g)(x) = \frac{(x+1)(x-1)}{x-1} = x + 1$. However, this doesn't hold when $x = 1$: $(f/g)(1) = (1^2 - 1)/(1 - 1) = 0/0$, which is undefined, and *not* equal to $1 + 1$. It's more accurate, therefore, to state that $(f/g)(x) = x + 1$ when $x \neq 1$, but is undefined when $x = 1$.

Example

Determine under what circumstances a relationship might be modeled by an arithmetic operation of two functions

A relationship might be modeled by an arithmetic operation of two functions if it represents a sum, product, difference, or quotient of two processes each modeled by a different function. For example, consider the following: A man has \$1000 in one bank account that generates 4% interest, compounded annually, and \$2000 in another bank account that generates 3% annual interest, continually compounded. We want to find the total amount of money he has in the bank after t years.

Applying the compound interest formula, $A = P\left(1 + \frac{r}{n}\right)^{nt}$, we find that the amount of money he has in the first account is equal to $f(t) = 1000(1.04)^t$. For continually compounded interest, the formula is $A = Pe^{rt}$, so the amount he has in the second account is equal to $g(t) = 2000e^{0.03t}$. The *total* amount of money he has in the bank is then simply the arithmetic sum of these two: $h(t) = f(t) + g(t) = 1000(1.04)^t + 2000e^{0.03t}$.

Example

Given two functions $f(x)$ and $g(x)$, show how you would determine the domain of each of the following

$(f+g)(x)$ $(f-g)(x)$ $(f \cdot g)(x)$ $(f/g)(x)$

In order for an arithmetic operation of two functions to be defined for a given x, each of the two functions themselves must also be defined at that x. In other words, in order to be in the domain of $(f+g)(x)$, $(f-g)(x)$, $(f \cdot g)(x)$, or $(f/g)(x)$, a particular number must be in the domain of both $f(x)$ and $g(x)$. For instance, if $f(x) = \ln(x+1)$ and $g(x) = \sqrt{2-x}$, since the domain of $f(x)$ is $(-1, \infty)$ and the domain of $g(x)$ is $(-\infty, 2]$, the domain of $(f+g)(x)$ is the intersection of those two domains, namely $(-1,2]$. For $(f/g)(x)$, we also have to take into account the fact that we can't divide by zero. Therefore, any values x for which $g(x) = 0$ are excluded from the domain of $(f/g)(x)$. For the sample $f(x)$ and $g(x)$ given above, for instance, the domain of $(f/g)(x)$ would be not $(-1,2]$ but $(-1,2)$, since $g(2) = 0$.

FUNCTION COMPOSITION

Function composition is a process in which the "output" of one function is used as the "input" of another. The composition of two functions $f(x)$ and $g(x)$ can be written as $(f \circ g)(x)$ or as $f(g(x))$.

HOW TO EVALUATE THE COMPOSITION OF TWO FUNCTIONS

This means we find the value of $f(x')$ at $x' = g(x)$. In other words, to evaluate the function composition $(f \circ g)(x)$ at $x = x_0$, we first find $y_0 = g(x_0)$ and then find $f(y_0)$, which is our final answer. For example, if $f(x) = x^2 + 1$ and $g(x) = x - 1$, then to find $(f \circ g)(3)$, we first find $g(3) = 2$, and then find $f(2) = 5$. So $(f \circ g)(3) = 5$. We can also rewrite the composition of two functions explicitly as a single function by substituting the expression for one function into the other. For instance, for the above functions, $(f \circ g)(x) = f(x-1) = (x-1)^2 + 1 = x^2 - 2x + 2$.

Example

Suppose $f(x) = x^2 - x$ and $g(x) = \frac{x}{x+1}$. How you would find each of the following

$(f \circ g)(3)$ $\quad g(f(-1))$ $\quad (f \circ f)(2)$ $\quad f\left(g(f(1))\right)$

All of these expressions indicate function composition, though they use different notations: the notation $(f \circ g)(x)$, for instance, is equivalent to $f(g(x))$. To evaluate a function composition, we can first evaluate the "inner" function expression, and then put that in as the input to the "outer" function. For instance, to evaluate $(f \circ g)(3)$, we first evaluate $g(3) = \frac{3}{3+1} = \frac{3}{4}$, and then $f(g(3)) = f\left(\frac{3}{4}\right) = \left(\frac{3}{4}\right)^2 - \frac{3}{4} = \frac{9}{16} - \frac{12}{16} = -3/16$. Similarly, $g(f(-1)) = g((-1)^2 - (-1)) = g(1+1) = g(2) = \frac{2}{2+1} = 2/3$, and $f(f(2)) = f(2^2 - 2) = f(2) = 2^2 - 2 = 2$. The last expression here can be evaluated similarly; it only takes an extra step: $f\left(g(f(1))\right) = f(g(1^2 - 1)) = f(g(0)) = f\left(\frac{0}{0+1}\right) = f(0) = 0^2 - 0 = 0$. In principle, we could apply the same procedure to a composition of any number of functions, such as $(f \circ g \circ g \circ f \circ g)(4)$.

Example

Suppose $f(x) = x^2 + 1$ and $g(x) = x - 1$. How would you write each of the following as a single function?

$f(g(x))$ $(g \circ f)(x)$ $(g \circ g)(x)$ $g\left(f(g(x))\right)$

All of these expressions indicate function composition, though they use different notations: the notation $(g \circ f)(x)$, for instance, is equivalent to $g(f(x))$. To rewrite a function composition as a single expression, we substitute the expression for one function in for x in the expression for the other. To find $f(g(x))$, for instance, we replace each x in the expression for $f(x)$—that is, $x^2 + 1$—with the expression for $g(x)$ —$(x - 1)$. This gives us $f(g(x)) = f(x - 1) = (x - 1)^2 + 1 = x^2 - 2x + 1 + 1 = x^2 - 2x + 2$. Similarly, $(g \circ f)(x) = g(f(x)) = (x^2 + 1) - 1 = x^2$, and $(g \circ g)(x) = g(g(x)) = (x - 1) - 1 = x - 2$.

The final expression, $g\left(f(g(x))\right)$, looks complex but requires the same procedure, just done in two steps. We've already found $f(g(x)) = x^2 - 2x + 2$. To find $g\left(f(g(x))\right)$, then, we just substitute this into the expression for $g(x)$: $g\left(f(g(x))\right) = g(x^2 - 2x + 2) = (x^2 - 2x + 2) - 1 = x^2 - 2x + 1$.

Under What Circumstances a Relationship Might Be Modeled by a Composition of Two Functions

A relationship might be modeled by a composition of two functions if it represents a process that depends on the output of another process. For example, consider the following: The population of a town increases exponentially according to the formula $P(t) = 1500e^{0.2t}$, where t is the time in years from a fixed starting year. On average, each person in the town eats 800 kilograms of food per year. We want to find the total amount of food eaten by the entire population of the town in year t.

The relationship between the total amount of food eaten in a year and the population of the town is straightforward: $f(x) = 800x$, where x equals the population. However, x is itself a function: specifically, it's $P(t)$, as given above. Therefore, the total amount of food, as a function of t, is $f(P(t)) = f(1500e^{0.2t}) = 800 \cdot 1500e^{0.2t} = 1200000e^{0.2t}$.

Example

Given two functions $f(x)$ and $g(x)$, show how you would determine the domain and range of $(f \circ g)(x)$

The notation $(f \circ g)(x)$ indicates function composition; this can also be written as $f(g(x))$. $(f \circ g)(x)$ can be evaluated at a point $x = x_0$ by first finding $g(x_0)$ and then putting that as the "input" for $f(x)$. In order for a value x_0 to be in the domain of $(f \circ g)(x)$, then, it's necessary first that x_0 be in the domain of $g(x)$ and second that $g(x_0)$ be in the domain of $f(x)$. For instance, if $f(x) = \sqrt{x}$ and $g(x) = x - 2$, then $x = 1$ is not in the domain of $(f \circ g)(x)$, even though it *is* in the domain of $f(x)$ and $g(x)$, since $g(1) = -1$ is not in the domain of $f(x)$.

Similarly, for a value $y = y_0$ to be in the range of $(f \circ g)(x)$, it's necessary not only that y_0 is in the range of $f(x)$, but that there is at least one value y' in the range of $g(x)$ such that $f(y') = y_0$.

Example

Given an arithmetic sequence expressed recursively, such as $a_1 = 2, a_n = a_{n-1} + 3$, explain how to write this sequence in an explicit (nonrecursive) form

An arithmetic sequence is one with a constant difference between consecutive terms: each term is equal to the preceding term plus (or minus) some constant number. An arithmetic sequence can be uniquely characterized by two specific numbers: the first element of the sequence a_1, and the sequence's common difference *d*. The latter is the difference between two consecutive terms of the sequence; *d* is positive if the sequence is increasing and negative if the sequence is decreasing.

An arithmetic sequence can be defined recursively by specifying a_1 and then giving the relationship between consecutive terms, $a_n = a_{n-1} + d$. It can also be defined explicitly with the expression $a_n = a_1 + d(n-1)$ (or, equivalently, $a_n = (a_1 - d) + dn$).

Here, we have a recursive definition; by inspecting it, we see that $a_1 = 2$ and $d = 3$. We can simply put those values into the explicit expression to get $a_n = 2 + 3(n-1)$ (or, equivalently, $a_n = -1 + 3n$).

Example

Given a geometric sequence expressed recursively, such as $a_1 = 3, a_n = 2a_{n-1}$, determine how to write this sequence in an explicit (nonrecursive) form

A geometric sequence is one with a constant ratio between consecutive terms: each term is equal to the preceding term multiplied (or divided) by some constant number. A geometric sequence can be uniquely characterized by two specific numbers: the first element of the sequence a_1, and the sequence's common ratio *r*. The latter is the ratio between two consecutive terms of the sequence; *r* is greater than one if the sequence is increasing and less than one if the sequence is decreasing. (If *r* is negative, the terms of the sequence alternate between positive and negative.)

A geometric sequence can be defined recursively by specifying a_1 and then giving the relationship between consecutive terms, $a_n = r \cdot a_{n-1}$. It can also be defined explicitly with the expression $a_n = a_1 \cdot r^{n-1}$ (or, equivalently, $a_n = (a_1/r) \cdot r^n$).

Here, we have a recursive definition; by inspecting it, we see that $a_1 = 3$ and $r = 2$. We can simply put those values into the explicit expression to get $a_n = 3 \cdot 2^{n-1}$ (or, equivalently, $a_n = \frac{3}{2} \cdot 2^n$).

Example

Given an arithmetic sequence defined explicitly, such as $a_n = 3 + 4n$, determine how you can write this sequence in a recursive form

An arithmetic sequence is a sequence in which each term differs from the previous amount by a fixed amount. An arithmetic sequence can be uniquely specified by this fixed difference, *d*, and the first term, a_1. The sequence can be expressed either explicitly, in the form $a_n = a_1 + d(n-1)$, or recursively, in the form $a_n = a_{n-1} + d$ (after specifying the first term).

In this case, being given the sequence in explicit form, we can find a_1 and d by inspection: $a_n = 3 + 4n = 3 + 4(1 + n - 1) = 3 + 4 + 4(n-1) = 7 + 4(n-1)$. So, comparing this to the general equation $a_n = a_1 + d(n-1)$, we find $a_1 = 7$ and $d = 4$. We can therefore write the sequence recursively as follows: $a_1 = 7, a_n = a_{n-1} + 4$.

Example

Given a geometric sequence defined explicitly, such as $a_n = 3\left(\frac{2}{3}\right)^n$,determine how you can write this sequence in a recursive form

A geometric sequence is a sequence in which each term differs from the previous amount by a fixed ratio. A geometric sequence can be uniquely specified by this fixed ratio, *r*, and the first term, a_1. The sequence can be expressed either explicitly, in the form $a_n = a_1 r^{n-1}$, or recursively, in the form $a_n = r \cdot a_{n-1}$ (after specifying the first term).

In this case, being given the sequence in explicit form, we can find a_1 and *r* by inspection: $a_n = 3\left(\frac{2}{3}\right)^n = 3\left(\frac{2}{3}\right)^{1+n-1} = 3\left(\frac{2}{3}\right)\left(\frac{2}{3}\right)^{n-1} = 2\left(\frac{2}{3}\right)^{n-1}$. So, comparing this to the general equation $a_n = a_1 r^{n-1}$, we find $a_1 = 2$ and r= $\frac{2}{3}$. We can therefore write the sequence recursively as follows: $a_1 = 2, a_n = \frac{2}{3}a_{n-1}$.

Situations Modeled by Arithmetic Sequences

An arithmetic sequence is a good model for any situation that involves a quantity that increases at equal intervals by a constant difference. That is, if the quantity is only defined relative to the positive integers (or technically at any other discrete, evenly spaced intervals), and the difference between the quantity corresponding to any two consecutive integers is the same, then it can be modeled by an arithmetic sequence.

Consider, for instance, the following situation: a large jar initially contains five marbles, and every day three more marbles are added to it. This fits the criteria described above: the quantity is defined relative to the positive integers (number of days), and the difference between the quantity corresponding to consecutive integers (the number of marbles on consecutive days) is constant (three). We can model this situation by the arithmetic sequence

$f(n) = 5 + 3(n-1)$.

Situations Modeled by Geometric Sequences

A geometric sequence is a good model for any situation that involves a quantity that increases at fixed intervals by a fixed ratio. That is, if the quantity is only defined relative to the positive integers (or technically at any other discrete, evenly spaced intervals), and the ratio of the quantity corresponding to any two consecutive integers is the same, then it can be modeled by a geometric sequence.

Consider, for instance, the following situation: the score for completing level one of a video game is 100, and the score for completing each level afterward is double the score for the previous level.

This fits the criteria described above: the quantity is defined relative to the positive integers (levels), and the ratio between the quantity corresponding to consecutive integers (the scores for completing two consecutive levels) is constant (two). We can model this situation by the geometric sequence $f(n) = 100 \cdot 2^{n-1}$.

Example

> Graphing of a function $f(x)$ compares with the graph of $f(x) + k$ for some constant k. What about the graph of $f(x + k)$
>
> If we replace $f(x)$ by $(x) + k$, this means the y coordinate of each point on the graph increases by k. This means the graph will be displaced vertically by a value of k: for instance, the graph of $y = x^2 + 1$ will be one unit higher than the graph of $y = x^2$. If k is negative, then the graph will be lowered.
>
> Replacing $f(x)$ by $f(x + k)$ effectively means the x coordinate of each point on the graph *de*creases by k. (Consider that $f\big((x - k) + k\big) = f(x)$, so evaluating $f(x)$ at $x = x_0$ yields the same value as evaluating $f(x + k)$ at $x = x_0 - k$.) This means the graph will be displaced horizontally to the left by a value of k: for instance, the graph of $y = (x + 1)^2$ will be one unit to the left of the graph of $y = x^2$. If k is negative, the graph will be displaced to the right.

Example

> Graphing of a function $f(x)$ compares with the graph of $kf(x)$ for some constant k. What about the graph of $f(kx)$
>
> If we replace $f(x)$ by $kf(x)$, this means the y coordinate of each point on the graph is multiplied k. This means the graph will be stretched vertically by a factor of k: for instance, the graph of $y = 2x^2$ will be stretched by a factor of 2 in the vertical direction relative to the graph of $y = x^2$. If k is less than one, then the graph will be compressed; if k is negative, it will be inverted vertically (in addition to the stretch or compression).
>
> Replacing $f(x)$ by $f(kx)$ effectively means the x coordinate of each point on the graph is *divided* by a factor of k. This means the graph will be compressed horizontally by a factor of k (or, if $k < 1$, stretched horizontally by a factor of $1/k$): for instance, the graph of $y = (2x)^2$ will be compressed horizontally by a factor of two relative to the graph of $y = x^2$. If k is negative, the graph will be inverted horizontally (in addition to the stretch or compression).

Determining from Inspection of a Graph Whether the Graph Represents an Odd or Even Function

An *odd function* is one for which $f(-x) = -f(x)$. This means that when the x coordinate is inverted, so is the y coordinate. The left half of the graph (to the left of the y-axis) will therefore be an inverted version of the right half (to the right of the y-axis). Rotating the graph by 180° inverts both the x and the y coordinates, so an odd function is symmetrical with respect to 180° rotations around the origin.

An *even function* is one for which $f(-x) = f(x)$. This means that when the x coordinate is inverted, the y coordinate does not change. The left half of the graph is therefore a mirror image of the right half. Reflecting the graph through the y-axis inverts the x coordinate without changing the y coordinate, so an even function is symmetrical with respect to reflections through the y-axis.

A graph that exhibits neither symmetry is neither even nor odd.

DETERMINING WHETHER A FUNCTION GIVEN ALGEBRAICALLY IS EVEN OR ODD

A function $f(x)$ is *even* if it is unchanged when x is replaced by $-x$. It is *odd* if the sign of the function is changed when x is replaced by $-x$. In other words, for an even function, $f(-x) = f(x)$; for an odd function, $f(-x) = -f(x)$. We can therefore check whether a function is even or odd by evaluating $f(-x)$ and checking whether (perhaps after some suitable algebraic manipulation) it is equal to $f(x)$ or $-f(x)$.

For instance, consider the function $f(x) = x^2 + 1$. $f(-x) = (-x)^2 + 1 = x^2 + 1 = f(x)$. $f(-x) = f(x)$, so this function is even. Now consider $g(x) = x^3 + \frac{1}{x}$. $g(-x) = (-x)^3 + \frac{1}{-x} = -x^3 - \frac{1}{x} = -\left(x^3 + \frac{1}{x}\right) = -g(x)$. $g(-x) = -g(x)$, so this function is odd. Finally, consider $h(x) = x - 1$. $h(-x) = -x - 1 = -(x + 1)$. This is equal to neither $h(x)$ nor $-h(x)$, so this function is neither even nor odd.

<u>Example</u>

> Given an expression for a function such as $f(x) = x^3 + 1$ or $g(x) = \frac{x}{x-1}$, determine how to find the function's inverse
>
> One algorithm for finding the inverse of the function is as follows: Solve the equation of the function for x—that is, manipulate the equation to get x by itself on one side of the equation. This equation now shows the inverse function of $f(x)$; to write it explicitly as an inverse, change the x to $f^{-1}(x)$ and change each instance of $f(x)$ on the other side of the equation to x.
>
> For instance, we can rewrite our first equation, $f(x) = x^3 + 1$, as $x^3 = f(x) - 1$, so $x = \sqrt[3]{f(x) - 1}$. Writing this explicitly as an inverse function, we can change the x to $f^{-1}(x)$ and the $f(x)$ to x to get $f^{-1}(x) = \sqrt[3]{x - 1}$. Our second example, $g(x) = \frac{x}{x-1}$, takes a little more work: multiplying both sides by $x - 1$ gives $(x - 1)g(x) = x$, which we can rearrange to $xg(x) - x = g(x)$, and finally to $x = \frac{g(x)}{g(x)-1}$. Replacing the x by $g^{-1}(x)$ and the $g(x)$ by x yields $g^{-1}(x) = \frac{x}{x-1}$: this function is its own inverse.

<u>Example</u>

> Given a function such as $f(x) = 3x^3 - 4$, determine how you could solve an equation such as $f(x) = 20$
>
> To solve an equation such as $f(x) = 20$, we're not asked to evaluate the function. Rather, we're asked to find a value for x at which the function evaluates to a given value. While in some cases it may be possible to solve such an equation through trial and error, a more reliable method is to use inverse functions (if $f(x)$ is invertible, which is the case here). If we find the inverse $f^{-1}(x)$ of $f(x)$, then we can apply that inverse function to both sides, yielding $f^{-1}(f(x)) = f^{-1}(20)$, which is to say $x = f^{-1}(20)$.

For our sample equation, $f(x) = 3x^3 - 4$, we can find the inverse algebraically. If $y = 3x^3 - 4$, then $3x^3 = y + 4$, $x^3 = \frac{y+4}{3}$, and finally $x = \sqrt[3]{(y+4)/3}$. So $f^{-1}(x) = \sqrt[3]{(x+4)/3}$. We can now find $f^{-1}(20)$: $f^{-1}(20) = \sqrt[3]{(20+4)/3} = \sqrt[3]{24/3} = \sqrt[3]{8} = 2$.

UNDER WHAT CIRCUMSTANCES AN EQUATION OF THE FORM $f(x) = C$ WILL HAVE A UNIQUE SOLUTION

An equation of the form $f(x) = C$ may have one solution, multiple solutions, or no solution. For example, $x + 1 = 1$ has one solution; $x^2 = 1$ has two solutions; $\sin x = 1$ has infinitely many solutions; and $e^x + 2 = 1$ has no solutions.

In order for $f(x) = C$ to have any solutions at all, C must be in the range of $f(x)$. (Note that 1 is not in the range of $f(x) = e^x + 2$.) However, this doesn't guarantee that the solution is unique. If the function $f(x)$ is invertible, then $f(x) = C$ is guaranteed to have a unique solution for all values of C in the range of $f(x)$. If not, then for at least some values of C the solution of $f(x) = C$ will not be unique. For instance, if $f(x) = x^3 - x$, then $f(x) = 0$ has three solutions, since $f(0) = f(1) = f(-1) = 0$, but $f(x) = 6$ has the unique solution $x = 2$.

Example

Given two functions expressed algebraically, explain how to determine whether they are inverses

The simplest way to determine algebraically whether two functions are inverses is to evaluate their composition. If $(f \circ g)(x) = x$ and $(g \circ f)(x) = x$, then $g(x) = f^{-1}(x)$ and vice versa; these two functions are inverses.

For example, consider $f(x) = x^3 + 1$ and $g(x) = \sqrt[3]{x-1}$. $(f \circ g)(x) = f\left(\sqrt[3]{x-1}\right) = \left(\sqrt[3]{x-1}\right)^3 + 1 = (x-1) + 1 = x$, and $(g \circ f)(x) = g(x^3+1) = \sqrt[3]{(x^3+1) - 1} = \sqrt[3]{x^3} = x$.. These two functions, therefore, are inverses. Some functions are their own inverses. Consider $h(x) = \frac{x}{x-1}$: $(h \circ h)(x) = h\left(\frac{x}{x-1}\right) = \frac{x/(x-1)}{x/(x-1)-1} = \frac{x}{x-(x-1)} = \frac{x}{1} = x$. So $h^{-1}(x) = h(x)$.

Example

If $f(g(x)) = x$ for all x in the domain of $g(x)$, and $g(f(x)) = x$ for all x in the domain of $f(x)$, what does this imply about the relationship between $f(x)$ and $g(x)$

If $f(g(x)) = x$ and $g(f(x)) = x$ for all x in the appropriate domains, then this means that $f(x)$ and $g(x)$ are inverse functions, i.e. $f^{-1}(x) = g(x)$ and $g^{-1}(x) = f(x)$. This means that if $f(x_0) = y_0$, then $g(y_0) = x_0$, and vice versa.

Inverse functions are useful because they can be used to solve equations of the form $f(x) = C$. We can get the solution by simply taking the inverse function of both sides: $f^{-1}(f(x)) = f^{-1}(C)$, but $f^{-1}(f(x)) = x$, so $x = f^{-1}(C)$. For example, the exponential function e^x and the logarithm $\ln x$ are inverses, so we can use the exponential to solve logarithmic functions, and vice versa: if $e^x = 3$, then $\ln e^x = \ln 3$, so $x = \ln 3$.

Example

Given a function represented by a table like the one below, determine how you can find the function's inverse

x	0	1	2	3	4	5	6	7
$f(x)$	4	5	0	2	7	1	6	3

To change a function into its inverse, essentially all we need to do is to interchange its domain and range. Therefore, when the function is given in table form, we can just switch the two rows, as follows:

x	4	5	0	2	7	1	6	3
$f^{-1}(x)$	0	1	2	3	4	5	6	7

Putting the domain in ascending order (which isn't technically necessary, but makes the table easier to refer to) gives:

X	0	1	2	3	4	5	6	7
$f^{-1}(x)$	2	5	3	7	0	1	6	4

So, for instance, $f^{-1}(0) = 2$, and $f^{-1}(1) = 5$ —as we would expect, since $f(2) = 0$ and $f(5) = 1$.

DESCRIBE HOW THE GRAPH OF A FUNCTION COMPARES TO THE GRAPH OF ITS INVERSE

To convert an invertible function $f(x)$ into its inverse, we interchange the domain and range. This means that the point (x, y) in $f(x)$ maps into the point (y, x) in $f^{-1}(x)$. On a graph, this is equivalent to reflecting the point through the line $y = x$, i.e. flipping it over diagonally, from upper left to lower right. Since this transformation applies to each point, it also applies to the function as a whole: we can get the graph of $f^{-1}(x)$ from the graph of $f(x)$ by just flipping the whole graph over through the diagonal line $y = x$. For example, if the graph of $f(x)$ is the graph shown on the left below, then flipping it over diagonally yields the graph on the right, which is the graph of $f^{-1}(x)$.

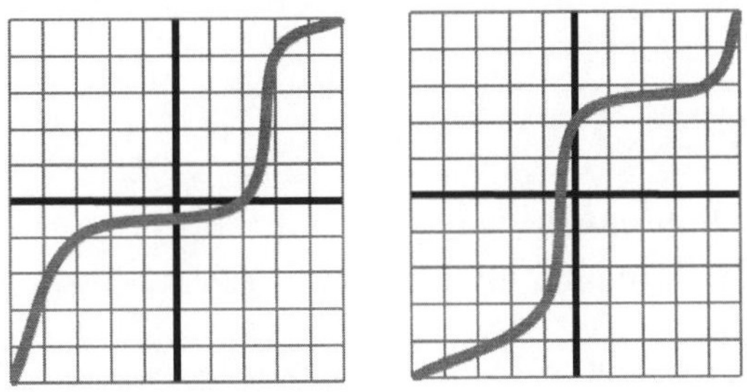

Example

Given a graph of an invertible function $f(x)$ and a value x_0, determine how you can read the value $f^{-1}(x_0)$ from the graph

While it's possible to turn the graph of $f(x)$ into a graph of $f^{-1}(x)$ by simply flipping it over diagonally through the line $y = x$, it's also possible to read values of $f^{-1}(x)$ off the graph of $f(x)$ directly. Just as we can read the value of $f(x_0)$ from a graph by finding the point where the graph intersects the line $x = x_0$ and observing the y coordinate of that point, so we can read the value of $f^{-1}(x_0)$ by the inverse process: We can find the point where the graph intersects the line $y = x_0$, and then read off the point's x-coordinate. The diagram below illustrates this procedure:

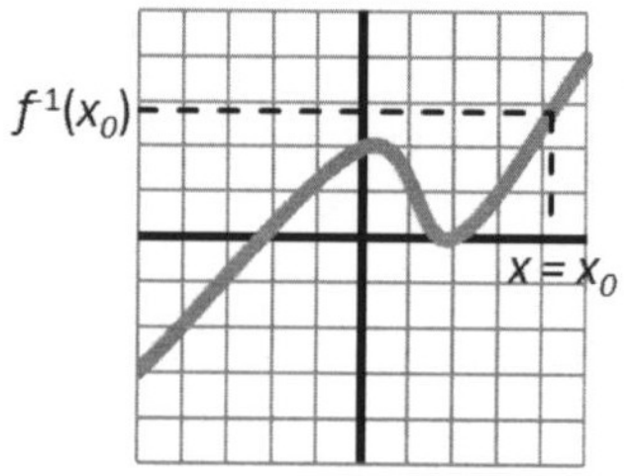

DETERMINE HOW YOU CAN TELL FROM THE GRAPH OF A FUNCTION WHETHER THE FUNCTION HAS AN INVERSE

A function has an inverse if and only if each point in the range corresponds to a unique point in the domain. This is similar to the definition of a function itself—in a function, each point in the domain corresponds to a unique point in the range. Therefore, we can use similar guidelines to identify invertible functions. One simple way to tell whether a graph represents a function is the "vertical line test"; we observe whether there exists any vertical line that intersects the graph in more than one point. Similarly, we can check whether a function is invertible from its graph by using a sort of "horizontal line test": we check whether there exists any *horizontal* line that intersects the function in more than one point. If so, then the y coordinate corresponding to that line is matched up with more than one x coordinate, which means that the function is *not* invertible. If no such horizontal line exists, then the function is invertible.

Example

Given a non-invertible function such as $f(x) = x^2 + 1$ or $g(x) = x^3 - x$, determine how you could produce an invertible function by restricting the domain

What makes these functions non-invertible is that there are certain y values that correspond to more than one x value. For instance, $f(1) = f(-1) = 2$, and $g(-1) = g(0) = g(1) = 0$. However, this only occurs because the graph is increasing in some parts of the domain and decreasing in others. If we restrict the domain to regions where the function is only increasing or decreasing, within these regions each y corresponds to a unique x, so this restricted function is invertible. Our first example, $f(x) = x^2 + 1$, decreases from $-\infty$ to 0, and increases from 0 to ∞. Restricting the domain to either the interval $(-\infty, 0]$ or $[0, \infty)$ produces an invertible function. Finding exactly where $g(x) = x^3 - x$ transitions between increasing and decreasing is difficult without calculus, but we don't need to know the precise transition points; we can choose a smaller interval over which it's clearly increasing or decreasing. For instance, if we graph $g(x)$, we can see it's clearly increasing for all $x > 1$, so we can get an invertible function by restricting its domain to $[1, \infty)$.

Example

Why is it incorrect (or at least incomplete) to say that the functions $f(x) = x^2$ and $g(x) = \sqrt{x}$ are inverses

At first, it might seem obvious that $f(x) = x^2$ and $g(x) = \sqrt{x}$ are inverses. After all, if we take the function composition $f(g(x))$, we find $f(g(x)) = f(\sqrt{x}) = (\sqrt{x})^2 = x$. However, one characteristic of inverse functions is that the domain of the inverse function is the same as the range of the original function, and vice versa. In this case, the range of $f(x)$ is indeed equal to the domain of $g(x)$: $[0, \infty)$. But the domain of $f(x)$ is not equal to the range of $g(x)$: the former is $(-\infty, \infty)$ but the latter only $[0, \infty)$. And, indeed, while $f(g(x)) = x$, we can find points for which $g(f(x)) \neq x$. For example, $g(f(-1)) = g((-1)^2) = g(1) = \sqrt{1} = 1$.

The root of the problem is that the function $f(x)$ is not invertible. There are elements in the range that match up to multiple points in the domain, such as $f(1) = f(-1) = 1$. If we make a new function $f_+(x)$ by restricting the domain of $f(x)$ to the interval $[0, \infty)$, then indeed $g(x) = f_+^{-1}(x)$.

Example

Is it possible that $f(g(x)) = x$ for all x in the domain of $g(x)$, but $g(f(x)) \neq x$ for some x in the domain of $f(x)$

Yes, it's possible that $f(g(x)) = x$ for all x in the domain of $g(x)$ without the same being true for $g(f(x))$. A simple example is $f(x) = x^2$ and $g(x) = \sqrt{x}$. For all x in the domain of $g(x)$, $f(g(x)) = f(\sqrt{x}) = (\sqrt{x})^2 = x$. However, this isn't true for $g(f(x))$: consider $g(f(-1)) = g((-1)^2) = g(1) = \sqrt{1} = 1 \neq -1$.

The reason that occurs is because in this case $f(x)$ is not an invertible function; there are multiple values of x that correspond to the same value of $f(x)$. Accordingly, there can be no function for which $f(g(x)) = x$ and $g(f(x)) = x$, because then the functions would be inverses. If we restrict $f(x)$ to the interval $[0, \infty)$, to produce a new function $f_2(x)$, then $f_2(x)$ is invertible, $f_2(g(x)) = x$, and $g(f_2(x)) = x$.

Example

Given a logarithmic equation such as $3\log_2 2x = 15$, determine how you could solve for x.

To solve a logarithmic equation where the variable appears in the argument of the logarithm, we have to use the fact that the exponential function is the inverse of the logarithmic. First, we rearrange the expression to get the logarithmic expression on a side by itself. After that, we can cancel the logarithm by taking the exponential function of both sides (with the same base as the logarithm). From that point on, solving the equation is a matter of simple algebra.

For instance, we can divide by three both sides of our sample equation, $3\log_2 2x = 15$, to get $\log_2 2x = \frac{15}{3} = 5$. We can now cancel the logarithm with an exponential: $2^{\log_2 2x} = 2^5$, or, simplifying the left hand side, $2x = 2^5$. So $x = \frac{2^5}{2} = \frac{32}{2} = 16$.

Relationship Between Exponential Functions and Logarithms

Exponential functions and logarithms are inverse functions. More specifically, an exponential function of a particular base is the inverse of a logarithmic function of the same base. For instance, e^x and $\ln x$ are inverse functions. 10^x and $\log_{10} x$ are inverse functions. And, in general, a^x and $\log_a x$ are inverse functions.

<u>How we can use this relationship to solve problems</u>

This relationship is often useful for solving problems, because we can use it to cancel an exponential or logarithmic function on one side of an equation. For example, given an equation of the form $a^x = C$, we can solve for x by taking the logarithm of both sides: $\log_a(a^x) = \log_a C$, which, since a^x and $\log_a x$ are inverse functions, reduces to $x = \log_a C$. Conversely, given an equation of the form $\log_a x = C$, we can solve for x by taking the exponential of both sides: $a^{\log_a x} = a^C$, which reduces to $x = a^C$.

Describe Under What Circumstances You Would Use a Logarithmic Function to Solve a Problem

The logarithm is the inverse of the exponential, and so can be used to cancel an exponential function. This is useful if a problem involves an exponential equation in which the variable to be solved for is in the exponent; we can then use the logarithmic equation to get the variable out of the exponent. For instance, we can solve the equation $10^x = 20$ by simply taking the logarithm, base 10, of both sides: $\log_{10}(10^x) = \log_{10}(20)$, thus $x = \log_{10} 20$.

More complex equations may require some algebraic manipulation: from $2e^{3x+5} + 6 = 20$, we get $2e^{3x+5} = 14$, then $e^{3x+5} = 7$. We can now take the logarithm of both sides: $\ln(e^{3x+5}) = \ln 7$, hence $3x + 5 = \ln 7$ and finally $x = (\ln 7 - 5)/3$.

<u>Example</u>

Describe what kind of function grows by equal differences over equal intervals, and prove your answer

A function that grows by equal differences over equal intervals is a *linear function*, a function of the form $f(x) = mx + b$.

To prove that a linear function meets this criterion, we can consider some arbitrary interval, Δx, and then show that $f(x + \Delta x) - f(x)$ does not depend on x—hence, that when x increases by an interval of Δx, $f(x)$ grows by a constant amount. Substituting in the linear function, we get$f(x + \Delta x) - f(x) = \big(m(x + \Delta x)\big) + b - (mx + b) = mx + m\Delta x + b - mx - b = m\Delta x$. As expected, this does not depend on x, proving that the linear function grows by equal differences over equal intervals.

<u>Example</u>

Describe what kind of function grows by equal factors over equal intervals, and prove your answer

A function that grows by equal factors over equal intervals is an *exponential function*, a function of the form $f(x) = Ae^{bx}$.

To prove that an exponential function meets this criterion, we can consider some arbitrary interval, Δx, and then show that $f(x + \Delta x)/f(x)$ does not depend on x—hence, that when x increases by an interval of Δx, $f(x)$ grows by a constant factor. Substituting in the exponential function, we get $f(x + \Delta x)/f(x) = Ae^{b(x+\Delta x)}/Ae^{bx} = Ae^{bx}e^{b\Delta x}/Ae^{bx} = e^{b\Delta x}$. As expected, this does not depend on x, proving that the exponential function grows by equal factors over equal intervals.

PHRASES IN A WORD PROBLEM INDICATING QUANTITY CHANGES

If one quantity changes at a constant rate relative to another, this means the two quantities have a linear relationship ($y = mx + b$). There are a number of common phrases in a word problem that might indicate this. Rarely are we told directly that the relationship between two quantities is linear, but there are other ways the relationship could be phrased. Consider the following statement: "For every meter you go deeper underwater, the pressure increases by ten thousand Pascals." This says that one quantity—here the pressure—increases at a constant rate—ten thousand Pascals per meter—relative to another—the depth.

Another common phrasing that indicates a linear relationship is the statement that one quantity is proportional to another. This can be expressed algebraically as $y = mx$. Note that not all linear relationships are proportional—only those in which the y intercept is zero—but all proportional relationships are linear.

Example

Describe what kind of equation you would use to model a relationship in which one quantity changes at a constant rate per unit interval relative to another

A relationship in which one quantity changes at a constant rate per unit interval relative to another is necessarily a *linear relationship*, and can be modeled by a linear equation, of the form $f(x) = mx + b$. In this equation, m is equal to the slope of the line, or to the rate at which $f(x)$ changes per unit interval. This rate may be given directly, or it may be possible to derive it from the given information: if, for example, two points in the relationship are given, we can find m by calculating $\frac{f(x_1)-f(x_2)}{x_1-x_2}$. The constant b is the y intercept, or the value $f(0)$; again, this may be given directly, or may be derivable from given information: if you know m and you're given one data point, you can put these known values into the equation $f(x) = mx + b$ and solve for b.

Example

Suppose you're told an event ticket costs a flat fee of $20 plus $5 per person. Describe how you would write an algebraic expression to describe the total price

To write an expression based on a given situation, the first thing to observe is what quantities we will be concerned with and that will correspond to variables in the expression. In this case, we have a total price that varies by the number of people involved, so we'll be writing $y = f(x)$, where y is the total cost and x is the number of people. Secondly, we can observe exactly how one quantity varies with the other. In this case, the price increases by a fixed amount for each additional person. A situation in which one quantity changes at a constant rate per unit interval relative to the other corresponds to a linear function. In this case, with a flat fee of $20 independent of the number of people, and a fee increase of $5 per person, an appropriate equation is $y = 20 + 5x$.

PHRASES IN A WORD PROBLEM INDICATING QUANTITY GROWS OR DECAYS

If one quantity grows or decays by a constant percent rate per unit interval relative to another, this means that the two quantities have an exponential relationship, $y = AB^{Cx}$—or rather, to take into account the percent rate, $y(t) = y_o(1 + P)^{Ct}$, where P is the percent rate. This relationship may be stated directly, but more often it must be partly inferred by other phrases appearing in the problem. For instance, we may be told the number of bacteria in a culture doubles every ten months—"doubles" indicates a percent rate of 100% ($1 + P = 2$, so $P = 1 = 100\%$).

Some common examples of problems involving exponential relationship involve compound interest and radioactive half-lives. In the former, the interest rate is the percent increase, while the coefficient of the exponent depends on how often the interest is compounded—for interest compounded monthly, the exponent would be $t/12$, assuming that t is measured in months. In the latter, the base of the exponent is always $\frac{1}{2}$, and the exponent is $t/t_{1/2}$, where $t_{1/2}$ is the half-life.

Example

Describe what kind of equation you would use to model a relationship in which one quantity changes by a constant percent rate per unit interval relative to another.

A relationship in which one quantity changes by a constant percent rate per unit interval relative to another is an *exponential relationship*, and can be modeled by an exponential equation, of the form $f(x) = Ae^{bx}$. If the percent rate of change is positive, then b is positive and the equation represents exponential growth; if the percent rate of change is negative, then b is negative and the equation represents exponential decay. Often in such a problem we are given the value of $f(x)$ when $x = 0$; this is simply A, as can be seen as follows: $f(0) = Ae^{b \cdot 0} = Ae^0 = A \cdot 1 = A$. Given this, if we have at least one other data point, we can use it to solve for b.

Example

Suppose you're told that an item initially sold for \$200 and its price has increased by 10% per year. Show how you would write an algebraic expression to describe the price

To write an expression based on a given situation, the first thing to observe is what quantities we will be concerned with and that will correspond to variables in the expression. In this case, we have a price that varies by year, so we'll be writing $y = f(t)$, where y is the total cost and t is the time in years from the item's initial sale. Secondly, we can observe exactly how one quantity varies with the other. In this case, the price increases by a fixed percentage per year. A situation in which one quantity changes at a constant percent rate per unit interval relative to the other corresponds to an exponential function. In this case, with an initial cost of \$200 at $t = 0$ and an increase of 10% per year, an appropriate equation is $y = 200 \cdot 1.1^t$.

Example

Given two points, such as (1,2) and (3,–1), describe how you can find a linear function that includes those points

A linear function is a function of the form $y = mx + b$, where m is the slope of the line and b is its y intercept. Suppose we are given two points, (x_1, y_1) and (x_2, y_2). We can find m by calculating the "rise over run": that is, $m = \frac{y_2 - y_1}{x_2 - x_1}$. Knowing m, we can then put the x and y values of one of the points into the equation $y = mx + b$ to solve for b, the only remaining unknown.

For example, using the sample points (1,2) and (3,–1) we have $m = \frac{-1-2}{3-1} = -\frac{3}{2}$. We can now put that into the linear equation along with the coordinates of the first point, (1,2), to yield $2 = -\frac{3}{2}(1) + b$, which we can solve for b to get $b = \frac{7}{2}$. (Had we used the second point instead of the first, we would have arrived at the same answer.) Therefore, the linear function that includes the two given points is $y = -\frac{3}{2}x + \frac{7}{2}$.

Example

Given two points, one of which has an x value of zero, such as (0,2) and (3,20), show how you can find an exponential function that includes those points

An exponential function has the form $y = Ae^{bx}$. (We could use a different base for the exponent rather than e, but in that case, by the properties of exponents, we'd simply have a different value for b; the form of the equation wouldn't change, and neither would the means of solving it.) If one of the given points has an x value of zero, then finding A is easy: we can just put the x and y values of this point into the equation, and we'll find that $y = A$. For the sample points, using (0,2) we find $2 = Ae^{b \cdot 0} = A$, so $A = 2$.

Knowing A, we can now use the x and y values of the other point to solve for b. For the sample points, putting in (3,20) we get $20 = 2e^{b \cdot 3}$. Dividing both sides by 2 yields $10 = e^{3b}$, and now we can take the natural logarithm of both sides, giving

$\ln 10 = 3b$, thus $b = \frac{\ln 10}{3}$. Therefore, our final equation is $y = 2e^{\frac{\ln 10}{3}x} = 2(e^{\ln 10})^{x/3} = 2 \cdot 10^{\frac{x}{3}}$.

Example

Given two points with nonzero *x* values, such as (2,4) and (6,60), show how you can find an exponential function that includes both point.

An exponential function has the form $y = Ae^{bx}$. Putting the *x* and *y* values for the given points into this equation yields two equations: in the case of the sample points, $4 = Ae^{2b}$ and $60 = Ae^{6b}$. We have two equations and two unknowns (A and b), so in principle it should be possible to solve for the unknowns.

There are many ways of doing this. We could, for instance, solve for A in one equation and substitute that value into the other. Another relatively simple method is to take the ratio of the two equations, thus canceling A: $\frac{60}{4} = \frac{Ae^{6b}}{Ae^{2b}}$, thus $15 = \frac{e^{6b}}{e^{2b}} = e^{6b-2b} = e^{4b}$. Taking the natural logarithm of both sides, $\ln 15 = 4b$, so $b = \frac{\ln 15}{4}$.

Knowing b, we can now use either point to solve for A: using (2,4), we get $4 = Ae^{\frac{\ln 15}{4} \cdot 2} = Ae^{\frac{\ln 15}{2}} = A(e^{\ln 15})^{\frac{1}{2}} = A \cdot 15^{1/2} = \sqrt{15}A$, so $A = \frac{4\sqrt{15}}{15}$. Our full equation is then $y = \frac{4\sqrt{15}}{15}e^{\frac{\ln 15}{4}x} = \frac{4\sqrt{15}}{15}(e^{\ln 15})^{x/4} = \frac{4\sqrt{15}}{15} \cdot 15^{\frac{x}{4}}$.

Example

Given a graph of a linear function, determine how you can construct an algebraic expression for the function

An algebraic expression for a linear function has the form $y = mx + b$. We can find both m and b from the graph of the function. b is simply equal to the *y* intercept, and can be read from the graph by observing at what point the graph of the linear function intersects the *y*-axis. To find m, the slope, we can take from the graph any two points that lie on the line, (x_1, y_1) and (x_2, y_2). The slope m is then equal to $m = \frac{y_1 - y_2}{x_1 - x_2}$.

There are two special cases that must be treated differently: horizontal and vertical lines. A horizontal line has a slope of zero; its equation is $y = b$, where b is, again, the *y* intercept (or the *y* coordinate of any other point on the line). Conversely, the slope of a vertical line is undefined; its equation is $x = a$, where a is the *x* intercept (or the *x* coordinate of any other point on the line).

Example

Given a graph of an exponential function, describe how you can construct an algebraic expression for the function

An algebraic expression for an exponential function has the form $y = Ae^{bx}$. A can be found directly from the graph; it is simply the graph's y intercept. To see this, consider what happens when we evaluate the expression at $x = 0$: $y = Ae^{b \cdot 0} = Ae^0 = A \cdot 1 = A$. Once we know A, we can choose any other point (x_1, y_1) that lies on the graph of the function *except* the y intercept. We can then use this to solve for b: $y_1 = Ae^{bx_1}$, so $\frac{y_1}{A} = e^{bx_1}$, $\ln\left(\frac{y_1}{A}\right) = \ln\left(e^{bx_1}\right) = bx_1$, and finally $b = \frac{1}{x_1}\ln\left(\frac{y_1}{A}\right)$. For instance, if the y intercept is $(0, 2)$ and the graph also passes through the point $(5,4)$, then we have $4 = 2e^{b \cdot 5}$, hence $2 = e^{5b}$, $\ln 2 = 5b$, and $b = \frac{\ln 2}{5}$. So an algebraic expression for the function is $y = 2e^{\frac{\ln 2}{5}x}$. This can be simplified: $2e^{\frac{\ln 2}{5}x} = 2\left(e^{\ln 2}\right)^{\frac{1}{5}x} = 2 \cdot 2^{\frac{1}{5}x} = 2^{\frac{1}{5}x+1}$. So $y = 2 \cdot 2^{\frac{1}{5}x}$ and $y = 2^{\frac{1}{5}x+1}$ are both acceptable ways to write the expression.

Example

Describe how you can write an algebraic expression for an arithmetic sequence given two (not necessarily consecutive) terms of the sequence

An arithmetic sequence is a sequence in which there is a constant difference d between consecutive terms. It can be expressed by the formula $f(n) = a_1 + d(n - 1)$. Suppose we are given two terms of the sequence, $f(n_1)$ and $f(n_2)$. Since the difference between consecutive terms is d, the difference between terms Δn terms apart is simply $\Delta n \cdot d$. In this case, $\Delta n = (n_2 - n_1)$, so $f(n_2) - f(n_1) = (n_2 - n_1)d$, and $d = \frac{f(n_2)-f(n_1)}{n_2-n_1}$. Knowing d, we can use one of the given terms to solve for a_1: $f(n_1) = a_1 + d(n_1 - 1)$, so $a_1 = f(n_1) - d(n_1 - 1)$.

For example, suppose we are told that the third term of an arithmetic sequence is 5 and the seventh term is 17—that is, $f(3) = 5$ and $f(7) = 17$. Then $17 - 5 = (7 - 3)d$, hence $d = \frac{17-5}{7-3} = \frac{12}{4} = 3$. We can now solve for a_1: $5 = a_1 + 3(3 - 1)$, so $a_1 = 5 - 3(3 - 1) = 5 - 3(2) = -1$. So an expression for this arithmetic sequence is $f(n) = -1 + 3(n - 1)$.

Example

Describe how you can write an algebraic expression for a geometric sequence given two (not necessarily consecutive) terms of the sequence

A geometric sequence is a sequence in which there is a constant ratio r between consecutive terms. It can be expressed by the formula $f(n) = a_1 r^{n-1}$. Suppose we are given two terms of the sequence, $f(n_1)$ and $f(n_2)$. Since the ratio of consecutive terms is r, the ratio of terms Δn terms apart is simply $r^{\Delta n}$. In this case, $\Delta n = (n_2 - n_1)$, so $f(n_2)/f(n_1) = r^{n_2 - n_1}$, and $r = \left(f(n_2)/f(n_1)\right)^{1/(n_2 - n_1)}$. Knowing r, we can use one of the given terms to solve for a_1: $f(n_1) = a_1 r^{n_1 - 1}$, so $a_1 = f(n_1)/r^{n_1 - 1}$.

For example, suppose we are told that the ninth term of an arithmetic sequence is 40 and the fifteenth term is 5—that is, $f(9) = 40$ and $f(15) = 5$. Then $\frac{5}{40} = r^{15-9} = r^6$, hence $r = \left(\frac{5}{40}\right)^{1/6} = \left(\frac{1}{8}\right)^{1/6} = \frac{1}{8^{\frac{1}{6}}} = \frac{1}{\sqrt[6]{8}} = \frac{1}{\sqrt{2}} = \frac{\sqrt{2}}{2}$. We can now solve for a_1: $40 = a_1 \left(\frac{\sqrt{2}}{2}\right)^{9-1} = a_1 \left(\frac{\sqrt{2}}{2}\right)^{8} = a_1 \left(\frac{1}{16}\right)$, so $a_1 = 40/(1/16) = 40 \cdot 16 = 640$. So an expression for this arithmetic sequence is $f(n) = 640 \cdot \left(\frac{\sqrt{2}}{2}\right)^{n-1}$.

Exponential Functions (with a Positive Exponent) Eventually Exceeds Any Polynomial Function

An exponential function has the form $y = Ae^{bx}$. A polynomial function has the form $y = a_n x^n + a_{n-1} x^{n-1} + \cdots + a_1 x + a_0$. The exponential always exceeds the polynomial for sufficiently large *x*, regardless of the values of *A*, *b*, *n*, and $a_0 \dots a_n$.

One way to see this is to consider the rate of change. One characteristic of an exponential function is that over a fixed interval it changes by a constant ratio: when *x* increases by 1, Ae^{bx} increases by e^b. This isn't true of polynomials; for large *x*, when only the leading term of the polynomial is significant, when *x* increases by 1 the polynomial increases by a ratio of $\frac{(x+1)^n}{x^n} = \left(\frac{x+1}{x}\right)^n$, which *decreases* as *x* increases, converging at very large *x* to 1. Therefore, eventually the rate of change of the polynomial will drop below the rate of change of the exponential, and thereafter at some point the exponential will surpass the polynomial.

Example

The graph below shows a cubic function, an exponential function, a linear function, a logarithmic function, and a quadratic function

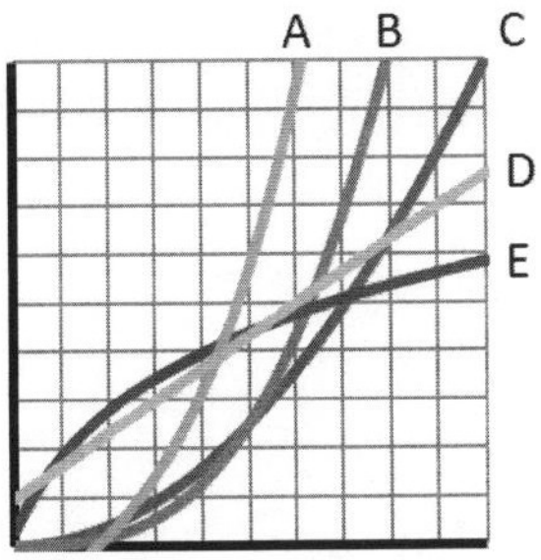

All of these functions tend toward infinity in their right end behavior, but some of them do so more rapidly than others. An exponential equation eventually increases more rapidly than any polynomial, and certainly increases more rapidly than a logarithmic function. The function that, in the end, increases most rapidly must therefore be the exponential, namely A. Conversely, the logarithmic function, the inverse of the exponential, must eventually increase more *slowly* than any polynomial. The logarithmic function must be the one that increases the most slowly in the end, namely E.

This leaves B, C, and D as the cubic, linear, and quadratic. These are all polynomial functions, and in its end behavior a polynomial of a larger order (largest exponent) increases more rapidly than one of a smaller order. This means B must be the cubic (order 3), C the quadratic (order 2), and D the linear (order 1).

Determining How to Tell Which Is a Cubic Function, an Exponential Function, a Linear Function, a Logarithmic Function, and a Quadratic Function

x	1	2	3	4	5	6	7	8	9	10
$f(x)$	0.00	0.60	1.60	3.00	4.80	7.00	9.60	12.60	16.00	19.80
$g(x)$	2.00	2.12	2.43	3.05	4.07	5.58	7.70	10.52	14.13	18.65
$h(x)$	4.00	5.39	6.20	6.77	7.22	7.58	7.89	8.16	8.39	8.61
$j(x)$	0.03	0.09	0.27	0.81	2.43	7.29	21.87	65.61	196.83	590.49
$k(x)$	0.50	1.00	1.50	2.00	2.50	3.00	3.50	4.00	4.50	5.00

$k(x)$ increases by a constant rate (0.5) per unit interval, and consecutive terms of $j(x)$ differ by a constant ratio (3), so $k(x)$ must be the linear function and $j(x)$ the exponential. However, we can also determine which function is which by comparing their end behavior. The function that, in the end, increases most rapidly must be the exponential, namely $j(x)$. Conversely, the logarithmic function, the inverse of the exponential, must eventually increase more *slowly* than any polynomial. Although $k(10) < h(10)$, it's clear that $h(x)$ is increasing at a much slower rate than $k(x)$, and for sufficiently large x in the end $k(x) > h(x)$. $h(x)$ is therefore the logarithmic function.

The remaining functions are all polynomials of varying degrees. In its end behavior a polynomial of a larger order (largest exponent) increases more rapidly than one of a smaller order. Observing the end behavior of $f(x)$, $g(x)$, and $k(x)$, we therefore conclude that $k(x)$ must be the lowest degree (linear), $f(x)$ the next (quadratic), and $g(x)$ the highest degree (cubic).

DETERMINING HOW TO EXPRESS THE SOLUTION OF AN EQUATION OF THE FORM $ab^{ct} = d$ IN TERMS OF A LOGARITHM

To express the solution of an equation of the form $ab^{ct} = d$ in terms of a logarithm, we'll have to take the logarithm of both sides of the equation. Before doing that, though, we'd have to isolate the exponential on a side by itself. We can do that by dividing both sides by a: $b^{ct} = d/a$. To cancel an exponential with a base of b, we have to use a logarithm with the base of b: $\log_b b^{ct} = \log_b(d/a)$. Since the logarithm is the inverse of the exponential, the left side, $\log_b b^{ct}$, reduces to just ct. So $ct = \log_b(d/a)$, so $t = \frac{\log_b(d/a)}{c}$. (Using the rule $\log(\alpha/\beta) = \log\alpha - \log\beta$, we can also rewrite this as $t = \frac{\log_b d - \log_b a}{c}$.)

Example

> Given an exponential equation such as $20 = 4 \cdot 2^{3x}$, show how you could solve for x
>
> To solve an exponential equation where the variable appears in the exponent, we have to use the fact that the logarithmic function is the inverse of the exponential. First, we rearrange the expression to get the exponential expression on a side by itself. After that, we can cancel the exponential by taking the logarithm of both sides (with the same base as the exponential). From that point on, solving the equation is a matter of simple algebra.
>
> For instance, we can divide by four both sides of our sample equation, $20 = 4 \cdot 2^{3x}$, to get $2^{3x} = \frac{20}{4} = 5$. We can now cancel the exponential with a logarithm: $\log_2 2^{3x} = \log_2 5$, or, simplifying the left hand side, $3x = \log_2 5$. So $x = \frac{1}{3}\log_2 5$.

USING A CALCULATOR TO EVALUATE LOGARITHMS SUCH AS $\log_e 2$, $\log_{10} 7$, AND $\log_2 5$

The logarithm base e is also called the natural logarithm, abbreviated *ln*: so $\log_e 2$ is the same thing as ln 2. Scientific calculators generally have an "ln" button, so evaluating natural logarithms is straightforward: to evaluate ln 2 on a TI-84 calculator, for instance, just involves hitting "LN", "2", "ENTER". Similarly, most scientific calculators have a button to calculate logarithms base 10, though it often just says "log" without the base. On a TI-84, $\log_{10} 7$ can be evaluated by hitting "LOG", "7", "ENTER".

Evaluating logarithms with bases other than 10 and e is not quite as straightforward, but still possible using the relation $\log_a b = \frac{\log_c b}{\log_c a}$ (for any base c). Therefore $\log_2 5 = \ln 5 / \ln 2$ or $\log_{10} 5 / \log_{10} 2$, either of which is readily evaluable with a calculator.

Example

> In the equation of a linear function, $y = mx + b$, what is the physical significance of the parameters m and b
>
> The parameter m is the slope of the line. Regarding the relationship between the physical quantities symbolized by x and y, m is the *rate of change* of y with respect to x: specifically, when x increases by 1, y increases by the amount m.
>
> b is often known as the *y-intercept* of the line: it represents the y coordinate of the point where the line crosses the y-axis. Another way of saying this, and one that

makes its physical significance clearer, is that b is the value of y when $x = 0$. When x represents time (as is frequently the case), then b is the *initial value* of y.

Example

Given an exponential equation such as the compound interest equation $A = P\left(1 + \frac{r}{n}\right)^{nt}$, determine how to interpret its parameters in terms of the context

While we could memorize the meaning of each parameter in the compound interest equation $A = P\left(1 + \frac{r}{n}\right)^{nt}$, it's also possible to figure out what they mean by examining the exponential equation. The amount of money in an account with compound interest rises exponentially as a function of time. The quantity that rises exponentially in this equation is A, so that must be the total amount of money in the account. The variable in the exponent is t, so that must be time. (There's also an n in the exponent, but that appears elsewhere as well.) P is the amount of money in the account when $t = 0$, so that must be the initial amount of money deposited in the account (the *principal*).

Two parameters of compound interest remain: the interest rate and the interval over which it's compounded (monthly, quarterly, etc.) The latter must be n, since it affects the exponent: interest compounded monthly would rise exponentially faster than interest compounded quarterly, for instance. That leaves r for the interest rate.

Example

Given a logarithmic equation such as the equation for the loudness of a sound in decibels in terms of its intensity, $\beta = 10\log_{10}\left(\frac{I}{I_0}\right)$, determine how to interpret its parameters in terms of the context

Even if we're not familiar with this equation, we can figure out what its parameters mean by examining the equation. Since the equation is for the loudness of a sound, that must be what the β stands for. This means the intensity, the other quantity mentioned, must be either I or I_0. If the intensity were I_0, that would mean that as the sound intensity increases, the loudness decreases, which seems illogical. It stands to reason, then that I represents the sound intensity. As for I_0, since it doesn't correspond to one of the variables, it must just be a constant factor.

As it happens, I_0 is defined as 10^{-12} watts, but even without knowing that, we can still get useful information from this equation; for instance, we can tell that if a sound's intensity increases by a factor of 100, then its loudness increases by 20 decibels.

Radian

A *radian* is a unit of angle measurement often used by mathematicians and scientists; one radian is equal to $180/\pi$ degrees. (The reason why radians are more useful than degrees may not be immediately obvious, but in calculus in particular, radians make calculations involving trigonometric functions much simpler.)

One way to define the radian measurement of an angle relates to the unit circle: if we place an angle with its vertex at the center of the circle, then the length of the arc subtended by the angle is equal to the measurement of the angle in radians. For example, a right angle subtends a quarter of the circle; the arc length is therefore one quarter of the circle's circumference, or $\frac{2\pi(1)}{4} = \frac{\pi}{2}$. So a right angle measures $\frac{\pi}{2}$ radians.

Converting an Angle between Angles and Radians, and Vice Versa

An angle's measurement in radians can be equated to the length of an arc of the unit circle subtended by an angle based at its center. Since the circumference of a full circle is 2π, this must correspond to 360°. The arc of a semicircle is just π, which must correspond to 180°. The arc of a quarter circle (the amount subtended by a right angle) is $\pi/2$, which must correspond to the number of degrees in a right angle, 90°. The conversion factor between degrees and radians must therefore be $\frac{2\pi}{360} = \frac{\pi/2}{90} = \frac{\pi}{180}$. So, for instance, 30° corresponds to $30\left(\frac{\pi}{180}\right) = \pi/6$ radians, 45° to $45\left(\frac{\pi}{180}\right) = \pi/4$ radians, and 60° to $60\left(\frac{\pi}{180}\right) = \pi/3$ radians. To convert from radians to degrees, we just *divide* by $\frac{\pi}{180}$, or multiply by $\frac{180}{\pi}$; thus, for instance, $\pi/5$ radians equals $(\pi/5)\left(\frac{180}{\pi}\right) = 36°$.

Making Sense of Trigonometric Functions of Angles Less than 0 or Greater than 90°

Trigonometric functions such as sine and cosine are often defined in terms of the ratio of sides of a right triangle. Obviously, a right triangle cannot have any angles less than 0 or greater than 90°. So there must be a more general definition of the trigonometric functions that doesn't rely on angles in a right triangle. The functions can instead be defined relative to a unit circle (a circle with radius 1), centered at the origin. If we start with a ray oriented along the positive *x* axis, and then rotate that ray by an angle θ counter-clockwise about the origin (or clockwise if θ is negative), then we can consider the point where that ray intersects the unit circle, defining cos θ to be equal to the *x* coordinate of this point and sin θ to be its *y* coordinate. For an angle between 0 and 90°, this definition coincides with the definition in terms of sides of a right triangle, but unlike that definition, the definition in terms of the unit circle allows us to make sense of trigonometric functions of any real number.

Trigonometric Functions Relation to the Unit Circle

To relate the trigonometric functions of an angle θ to the unit circle, first consider a ray that coincides with the *x*-axis, and then rotate that ray counterclockwise through the angle θ. Now, consider the *x* and *y* coordinates of the point where the rotated ray intersects a unit circle centered on the origin. (See the diagram below.)

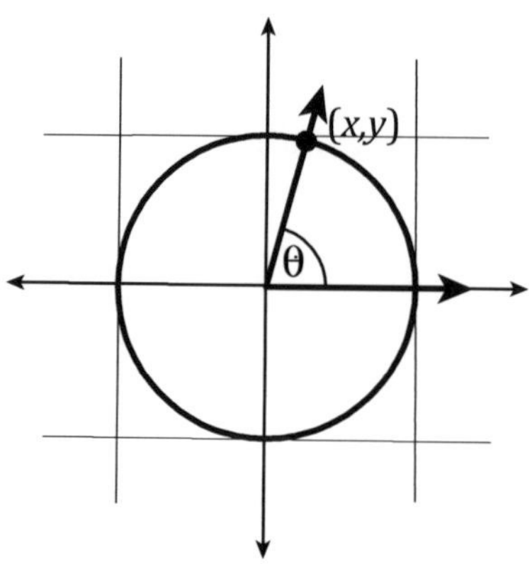

All six trigonometric functions can be described in terms of those coordinates. $\sin\theta = y$, $\cos\theta = x$, $\tan\theta = y/x$, $\sec\theta = 1/x$, $\csc\theta = 1/y$, and $\cot\theta = x/y$. If the denominator of any of these expressions is zero, then the corresponding function is undefined. For example, if $\theta = \pi/2$, then $x = 0$ and $y = 1$, so $\sin(\pi/2) = 1$, $\cos(\pi/2) = 0$, $\csc(\pi/2) = 1/1 = 1$, and $\cot(\pi/2) = 0/1 = 0$. Both $\tan(\pi/2)$ and $\sec(\pi/2)$ are undefined, since the expressions $1/x$ and y/x both have zeroes in their denominators.

Unit Circle vs. Right Triangles

Students first learn the trigonometric functions in terms of right triangles: the sine is equal to the length of the opposite leg divided by the length of the hypotenuse, the cosine to the length of the adjacent leg divided by the length of the hypotenuse, and the tangent to the length of the opposite leg divided by the length of the adjacent leg. While this method of determining the trigonometric functions remains often useful, it has its limitations. Namely, it only makes sense for angles between 0 and 90 degrees (0 and $\pi/2$ radians), since a right triangle can't have any angles with measurements outside this range. However, the trigonometric functions can also be defined in terms of the arc of a unit circle subtended by an angle at the circle's center. For angles between 0 and 90 degrees, this matches the right angle definition, but it also applies to angle measurements of any real number.

Finding the Sine, Cosine, and Tangent of $\pi - x$ in Terms of the Respective Functions of X

$\sin(\pi - x) = \sin(x)$, $\cos(\pi - x) = -\cos(x)$, and $\tan(\pi - x) = -\tan(x)$. While it's possible to memorize these relationships, it is unnecessary, since they can easily be derived from the unit circle. Consider the relationship between the angles θ and $\pi - \theta$ on the unit circle:

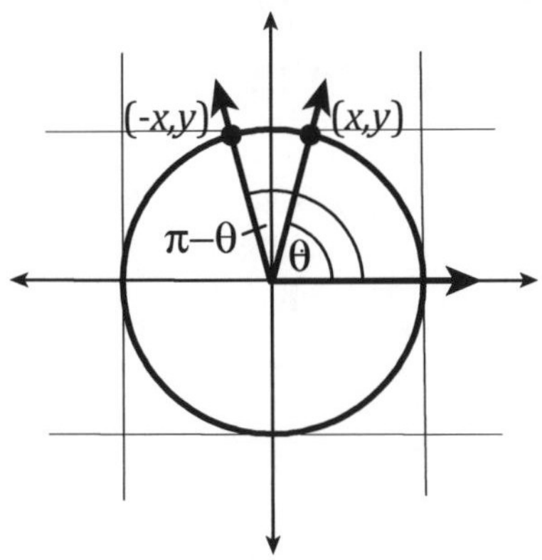

Replacing θ by $\pi - \theta$ is equivalent to a reflection through the y-axis: the x coordinate is inverted, and the y coordinate is unchanged. This means that the cosine is inverted (since it's equal to the x coordinate), the sine is unchanged (since it's equal to the y coordinate), and the tangent is inverted (since it's equal to the ratio y / x, and if one side of the ratio is inverted the ratio is inverted).

FINDING THE SINE, COSINE, AND TANGENT OF $\pi + x$ IN TERMS OF THE RESPECTIVE FUNCTIONS OF X

$\sin(\pi + x) = -\sin(x)$, $\cos(\pi + x) = -\cos(x)$, and $\tan(\pi + x) = \tan(x)$. While it's possible to memorize these relationships, it is unnecessary, since they can easily be derived from the unit circle. Consider the relationship between the angles θ and $\pi + \theta$ on the unit circle:

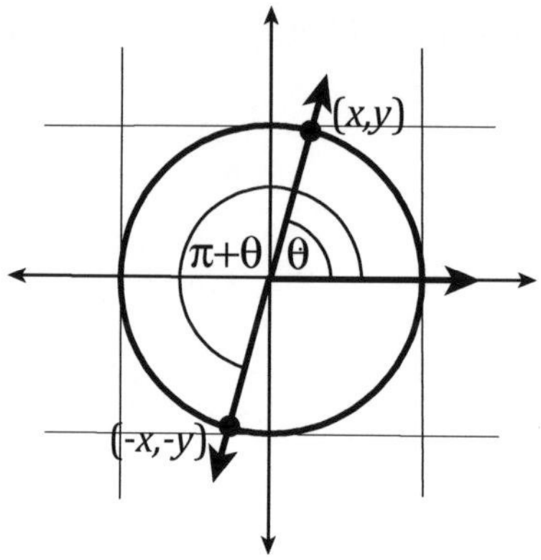

Replacing θ by $\pi + \theta$ means a rotation by π radians (or 180°), which is equivalent to reflections through both the *x* and *y* axes: the *x* and *y* coordinates are both inverted. This means that the cosine is inverted (since it's equal to the *x* coordinate), the sine is inverted (since it's equal to the *y* coordinate), and the tangent is unchanged (since it's equal to the ratio y / x, and the inversion of one side of the ratio cancels the inversion of the other).

FINDING THE SINE, COSINE, AND TANGENT OF 2π − X IN TERMS OF THE RESPECTIVE FUNCTIONS OF X

$\sin(2\pi - x) = -\sin(x)$, $\cos(2\pi - x) = \cos(x)$, and $\tan(2\pi - x) = -\tan(x)$. While it's possible to memorize these relationships, it is unnecessary, since they can easily be derived from the unit circle. Consider the relationship between the angles θ and $2\pi - \theta$ on the unit circle:

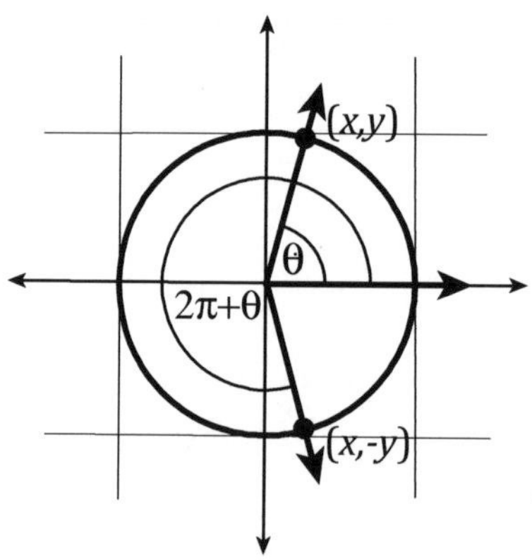

Replacing θ by $2\pi - \theta$ is equivalent to a reflection through the *x*-axis: the *y* coordinate is inverted, and the *x* coordinate is unchanged. This means that the cosine is unchanged (since it's equal to the *x* coordinate), the sine is inverted (since it's equal to the *y* coordinate), and the tangent is inverted (since it's equal to the ratio y / x, and if one side of the ratio is inverted the ratio is inverted).

FINDING THE SINE, COSINE, AND TANGENT OF $\pi/3$ WITHOUT USING TECHNOLOGY

The angle $\pi/3$ comes up often enough in problems that its trigonometric functions may be worth memorizing. However, if you don't memorize them, they're still not hard to find. If $\pi/3$ doesn't seem familiar, try converting it from radians to degrees: $\pi/3$ radians is equal to $(\pi/3)(180/\pi) = 60$ degrees. Now, consider a right triangle with one angle of 60°. This should sound familiar; a "30/60/90" triangle is one of the most familiar kinds of right triangles, and you should have learned that its hypotenuse is always twice as long as its shorter leg. Since the cosine of an angle in a right triangle is equal to the length of the adjacent leg divided by the length of the hypotenuse, and the adjacent leg to the 60° angle is the shorter leg, this means the cosine of 60°, and therefore of $\pi/3$, must be equal to ½. Therefore $\sin(\pi/3) = \sqrt{1 - (^1/_2)^2} = {}^{\sqrt{3}}/_2$, and $\tan(\pi/3) = (\sqrt{3}/2)/(1/2) = \sqrt{3}$.

Finding the Sine, Cosine, and Tangent of $\pi/6$ Without Using Technology

The angle $\pi/6$ comes up often enough in problems that its trigonometric functions may be worth memorizing. However, if you don't memorize them, they're still not hard to find. If $\pi/6$ doesn't seem familiar, try converting it from radians to degrees: $\pi/6$ radians is equal to $(\pi/6)(180/\pi) = 30$ degrees. Now, consider a right triangle with one angle of 30°. This should sound familiar; a "30/60/90" triangle is one of the most familiar kinds of right triangles, and you should have learned that its hypotenuse is always twice as long as its shorter leg. Since the sine of an angle in a right triangle is equal to the length of the opposite leg divided by the length of the hypotenuse, and the opposite leg to the 30° angle is the shorter leg, this means the sine of 30°, and therefore of $\pi/6$, must be equal to ½. Therefore $\cos(\pi/6) = \sqrt{1 - (^1/_2)^2} = {}^{\sqrt{3}}/_2$, and $\tan(\pi/6) = (1/2)/(\sqrt{3}/2) = \sqrt{3}/3$.

Finding the Sine, Cosine, and Tangent of $\pi/4$ Without Using Technology

The angle $\pi/4$ comes up often enough in problems that its trigonometric functions may be worth memorizing. However, if you don't memorize them, they're still not hard to find. If $\pi/4$ doesn't seem familiar, try converting it from radians to degrees: $\pi/4$ radians is equal to $(\pi/4)(180/\pi) = 45$ degrees. Now, consider a right triangle with one angle of 45°. This should sound familiar; a "45/45/90" triangle is one of the most familiar kinds of right triangles, and you should have learned that its legs are equal in length. This means if each leg has length 1, the hypotenuse, by the Pythagorean theorem, has length $\sqrt{1^2 + 1^2} = \sqrt{2}$. Since the sine of an angle in a right triangle is equal to the length of the opposite leg divided by the length of the hypotenuse, this means the sine of 45°, and therefore of $\pi/4$, must be equal to $1/\sqrt{2} = \sqrt{2}/2$. Similarly, $\cos(\pi/4) = 1/\sqrt{2} = \sqrt{2}/2$, and $\tan(\pi/4) = 1/1 = 1$.

Determining Which of the Six Trigonometric Functions Are Odd, and Which Are Even

An *odd* function is a function that is inverted when its argument is inverted: $f(x) = -f(-x)$. An *even* function is a function that is unchanged when its argument is inverted: $f(x) = f(-x)$. To see which trigonometric functions are odd and which are even, consider the unit circle: inverting the angle is equivalent to tracing the angle around the unit circle in the opposite direction.

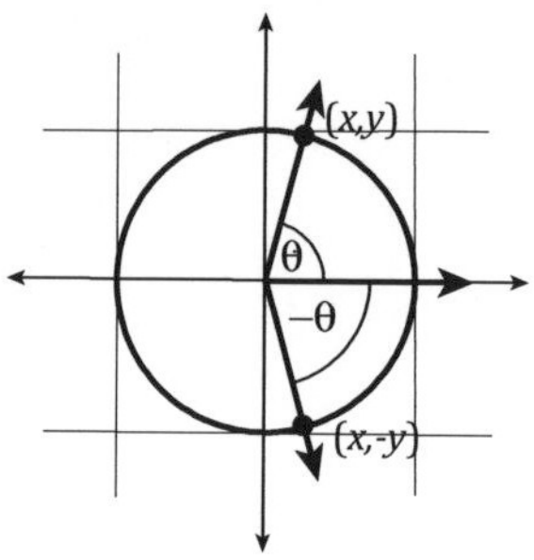

As is clear from the diagram, inverting the angle inverts the y coordinate (the sine) while leaving the x coordinate (the cosine) unchanged, so the sine is odd and the cosine even. The tangent, the ratio of the sine and cosine, is also inverted and therefore odd. If a function is odd or even, so is its reciprocal, so the cosecant and the cotangent are odd (as the reciprocals of the sine and tangent, respectively) and the secant is even (as the reciprocal of the cosine).

PERIODICITY OF THE SINE, COSINE, AND TANGENT FUNCTIONS

The sine and cosine functions are periodic with a period of 2π radians (360°): $\sin(\theta + 2\pi) = \sin(\theta)$, and $\cos(\theta + 2\pi) = \cos(\theta)$. This can be seen from their relation to the unit circle: since the sine and cosine of θ can be equated to the y and x coordinates of a point originally at (1,0) after it rotates by an angle θ about the origin, we would expect that after one full rotation—2π radians—the values would repeat.

The tangent is also periodic, but with a different period. Consider that the tangent is the ratio of the sine and the cosine. After a rotation of π radians—halfway around the unit circle—the sine and the cosine have both changed in sign: $\sin(\theta + \pi) = -\sin(\theta)$, and $\cos(\theta + \pi) = -\cos(\theta)$. But if both the sine and cosine change in sign, then their ratio is unchanged; the sign changes cancel out. Therefore $\tan(\theta + \pi) = \tan(\theta)$, and the period of the tangent is π.

Example

> What kind of function would best model periodic data such as the graph shown below

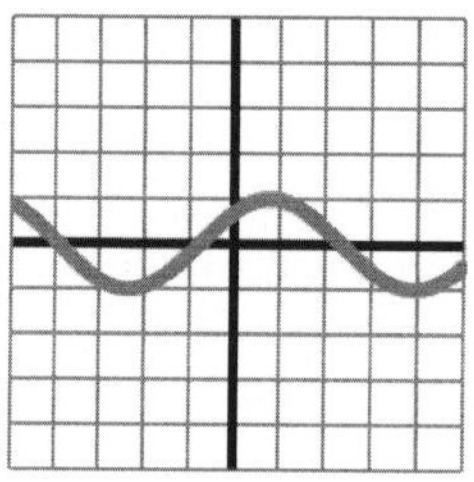

> A periodically oscillating graph like the one pictured here is best modeled by a sine wave or a cosine wave. Generally, if at $x = 0$ the wave is at its center and increasing, it's best modeled by $y = \sin(x)$ (or, more generally, $y = A(\sin bx) + C$, depending on the amplitude, period, and vertical displacement, but in the case shown here $A = b = 1$ and $C = 0$). If at $x = 0$ the wave is at its center and decreasing, the best model is $y = -\sin(x)$. If at $x = 0$ the graph is at its maximum, the appropriate function is

$y = \cos(x)$. If at $x = 0$ the graph is at its minimum, the appropriate function is $y = -\cos(x)$.

In a case like the one here where none of these apply, we require an appropriate *phase shift*. We can use either the sine or the cosine, since they differ only in phase; $\sin(x) = \cos\left(x + \frac{\pi}{2}\right)$. The particular graph shown here can be modeled by either $y = \sin\left(x + \frac{\pi}{4}\right)$ or $y = \cos\left(x - \frac{\pi}{4}\right)$; the two expressions are equivalent.

Writing a Trigonometric Equation to Represent a Sine Graph Like the One Shown Here

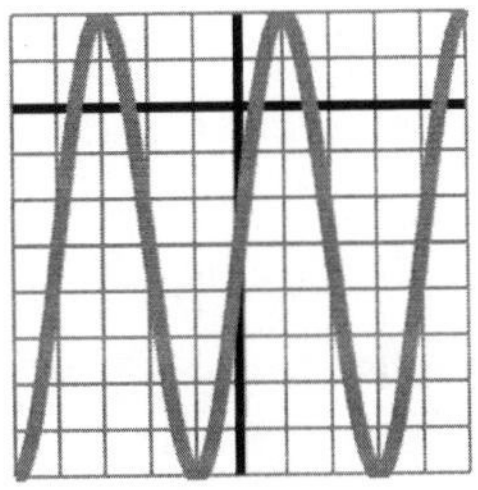

A sine function in its most general form takes the form $y = A\sin(bx + \phi) + C$. When the graph starts at the center line, as here, the phase shift ϕ is equal to zero. However, the other three parameters we still have to determine.

C here is the vertical displacement, that is, the distance between the center line of the graph and the x axis. In this case, the center line of the graph is 3 units below the x axis, so $C = -3$. A is the amplitude, the vertical distance between the center line and the maximum (or minimum) of the graph, or, equivalently, half the vertical distance between the maximum or minimum: in this case, $A = 5$. Finally, b is related to the period of the graph: more specifically, the period is equal to $^{2\pi}/_{b}$. In this case, the period (the horizontal distance between successive maxima or minima) is equal to 4, so $b = {}^{2\pi}/_{4} = \frac{\pi}{2}$. The full equation is therefore $y = 5\sin\left(\frac{\pi}{2}x\right) - 3$.

Example

Given the coordinates of an adjacent maximum and minimum of a periodic phenomenon, show how to determine the parameters of a trigonometric function that models the phenomenon

The fact that the phenomenon *has* a maximum and minimum implies it's best modeled by a sine or cosine, not a tangent. A sine or cosine function has four parameters: amplitude, period, vertical shift, and phase shift. The amplitude is half the difference between the y coordinates of the two points, while the period is twice the difference between the x coordinates. The vertical shift is the average of the y coordinates. As for the phase shift, it would be equal to the position of the maximum minus one quarter of the period.

For instance, suppose a function has a maximum at $(\pi/4, 5)$ and a minimum at $(7\pi/4, 1)$. The amplitude is then $\frac{5-1}{2} = 2$, the period is $2(7\pi/4 - \pi/4) = 3\pi$, the vertical shift is $\frac{5+1}{2} = 3$, and the phase shift is $\pi/4 - \frac{1}{4}(3\pi) = -\frac{\pi}{2}$.

Example

> Given that sin (x) is defined for any real value of x, explain why $\sin^{-1}(x)$ is always between $-\frac{\pi}{2}$ and $\frac{\pi}{2}$
>
> It's true that $y = \sin(x)$ is defined for any real number x, but each value of x does not correspond to a unique value of y. For instance, $\sin\left(\frac{\pi}{4}\right) = \sin\left(\frac{3\pi}{4}\right) = \sin\left(\frac{9\pi}{4}\right)$. Because of this, if we take the domain to include all real numbers, sin (x) is not invertible. To make a suitable invertible function, we have to restrict the domain to an interval where the function is strictly increasing or strictly decreasing. $\left[-\frac{\pi}{2}, \frac{\pi}{2}\right]$ is such an interval; between $-\frac{\pi}{2}$ and $\frac{\pi}{2}$ the function sin (x) is strictly increasing, and therefore invertible. It's not the *only* such interval—$\left[\frac{\pi}{2}, \frac{3\pi}{2}\right]$ or $\left[\frac{3\pi}{2}, \frac{5\pi}{2}\right]$ also qualify—but it makes sense to choose a domain that includes 0. Since inverting a function involves interchanging the range and the domain, if we restrict the original function to the domain $\left[-\frac{\pi}{2}, \frac{\pi}{2}\right]$, then this will be the range of the inverse function.

RANGES OF $\sin^{-1}(x)$, $\cos^{-1}(x)$, AND $\tan^{-1}(x)$

The trigonometric functions in their standard form are not invertible; there are multiple values of x that would give the same value for sin (x), cos (x), and tan (x). In order to produce invertible functions, we must restrict their domains to intervals in which they are strictly increasing or strictly decreasing. Although there are infinitely many such intervals we could choose for each function, it makes sense for simplicity's sake to choose continuous intervals that include zero and that include positive numbers. Such an interval for sin (x) is $\left[-\frac{\pi}{2}, \frac{\pi}{2}\right]$, for cos (x) is $[0, \pi]$, and for tan (x) is $\left[-\frac{\pi}{2}, \frac{\pi}{2}\right]$.

The range of an inverse function is equal to the domain of the original function. So the range of $\sin^{-1}(x)$ is $\left[-\frac{\pi}{2}, \frac{\pi}{2}\right]$, the range of $\cos^{-1}(x)$ is $[0, \pi]$, and the range of $\tan^{-1}(x)$ is $\left[-\frac{\pi}{2}, \frac{\pi}{2}\right]$.

Example

> Given an equation with an unknown in the argument of a sine or cosine function, such as $3 = 6\cos(2x)$, show how you can solve for the unknown
>
> To solve for an unknown in the argument of a trigonometric function, we have to cancel the function by using the appropriate inverse trigonometric function. We first manipulate the equation to get the trigonometric function on a side by itself; we can then take the corresponding inverse trigonometric function of both sides. For instance, for our sample equation, $3 = 6\cos(2x)$, we first divide by 6 to get $^1/_2 = \cos(2x)$, hence $2x = \cos^{-1}\left(\frac{1}{2}\right) = \frac{\pi}{3}$, and $x = \frac{\pi}{6}$.
>
> This is not, however, the only solution. Recall that $\cos(x) = \cos(x + 2\pi n) = \cos(-x + 2\pi n)$ for any integer n. The equation, therefore, has infinitely many solutions: not only can $2x$ equal $\frac{\pi}{3}$, but it can also be $\frac{\pi}{3} + 2\pi n$ or $-\frac{\pi}{3} + 2\pi n$, and so $x = \frac{1}{2}\left(\pm\frac{\pi}{3} + 2\pi n\right) = \pm\frac{\pi}{6} + \pi n$ for any integer n. Similar considerations apply for

other trigonometric functions: $\sin(x) = \sin(x + 2\pi n) = \sin(\pi - x + 2\pi n)$, and $\tan(x) = \tan(x + \pi n)$.

Example

Given an equation with an unknown in the argument of a tangent function, such as $\tan^2(x + 1) = 3$, show how you can solve for the unknown

To solve for an unknown in the argument of a trigonometric function, we have to cancel the function by using the appropriate inverse trigonometric function. We first manipulate the equation to get the trigonometric function on a side by itself; we can then take the corresponding inverse trigonometric function of both sides. For instance, for our sample equation, $\tan^2(x + 1) = 3$, we first take the square root of both sides to get $\tan\ (x + 1) = \pm\sqrt{3}$, hence $x + 1 = \tan^{-1}\sqrt{3} =$ or $\tan^{-1}(-\sqrt{3})$, and so $x + 1 = \pm\frac{\pi}{3}$.

This is not, however, the only solution. Recall that $\tan(x) = \tan(x + \pi n)$ for any integer n. The equation, therefore, has infinitely many solutions: not only can $x + 1$ equal $\pm\frac{\pi}{3}$, but it can also be $\pm\frac{\pi}{3} + \pi n$, and so $x = \pm\frac{\pi}{3} + \pi n - 1$ for any integer n. Similar considerations apply for other trigonometric functions: $\sin(x) = \sin(x + 2\pi n) = \sin(\pi - x + 2\pi n)$, and $\cos(x) = \cos(x + 2\pi n) = \cos(-x + 2\pi n)$.

PROVING THE PYTHAGOREAN IDENTITY, $\sin^2(\theta) + \cos^2(\theta) = 1$

There are many different ways to prove the Pythagorean identity. Probably the simplest method not using calculus or infinite series, however, relies on the well-known Pythagorean theorem that the sum of the squares of the legs of a right triangle is equal to the square of the hypotenuse. First of all, we can plot the angle on the unit circle, and then draw an appropriate triangle as shown below:

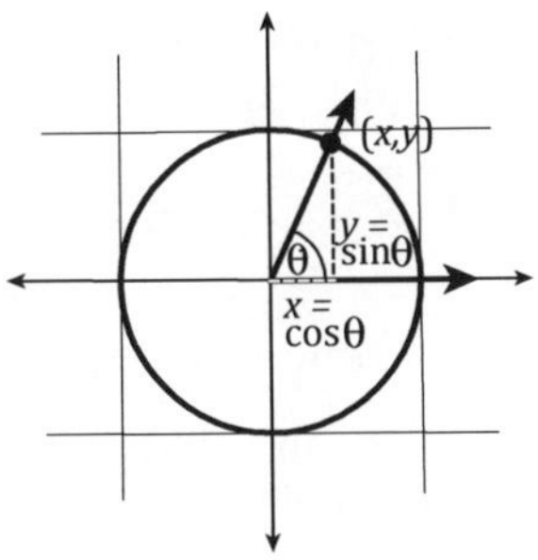

Notice that the hypotenuse is a radius of the circle, which by definition of a unit circle is equal to 1. The legs of the right triangle are equal to $|\sin\theta|$ and $|\cos\theta|$. The sum of the squares of the legs is then $\sin^2\theta + \cos^2\theta$, and the square of the hypotenuse is $1^2 = 1$. Thus, $\sin^2(\theta) + \cos^2(\theta) = 1$.

IF YOU'RE GIVEN THE VALUE OF ONE TRIGONOMETRIC FUNCTION OF AN ANGLE, WHAT OTHER INFORMATION YOU WOULD NEED IN ORDER TO FIND THE VALUE OF ANOTHER TRIGONOMETRIC FUNCTION OF THE SAME ANGLE

If the two trigonometric functions are reciprocals, no further information is needed. If we're given $\sin\theta$ and asked to find $\csc\theta$, or vice versa, all we have to do is take the reciprocal, since $\sin\theta = 1/\csc\theta$. The same goes for $\cos\theta$ and $\sec\theta$, or $\tan\theta$ and $\cot\theta$.

In other cases, however, the one function alone is insufficient. We can use the Pythagorean identities such as $\sin^2\theta + \cos^2\theta = 1$ or $1 + \cot^2\theta = \csc^2\theta$, but these identities alone will give us only the *absolute value* of the unknown function; it won't tell us whether it's positive or negative. For that, we need more information, such as the quadrant of the angle—then we can use the Pythagorean identities to find the magnitude of the function and the quadrant to determine the sign.

Example

If you're given the value of $\sin\theta$ and the quadrant of the angle, show how you can find $\cos\theta$ and $\tan\theta$

Given the value of $\sin\theta$, we can find $\cos\theta$ using the Pythagorean identity, $\sin^2(\theta) + \cos^2(\theta) = 1$. Solving for $\cos\theta$, we get $\cos^2\theta = 1 - \sin^2\theta$, and $\cos\theta = \pm\sqrt{1-\sin^2\theta}$. For instance, if $\sin\theta = \frac{3}{5}$, then $\cos\theta = \pm\sqrt{1-\left(\frac{3}{5}\right)^2} = \pm\sqrt{1-\frac{9}{25}} = \pm\sqrt{\frac{16}{25}} = \pm\frac{4}{5}$. To decide whether the cosine is positive or negative, we take into account the quadrant of the angle: if the angle is in quadrant 1 or 4, $\cos\theta$ is positive, and if the angle is in quadrant 2 or 3, $\cos\theta$ is negative.

Since the tangent is equal to the sine divided by the cosine, once we know the sine and the cosine finding the tangent is simple. Again, if $\sin\theta = \frac{3}{5}$, and if the angle is in quadrant 2 so the cosine is negative, then $\tan\theta = \frac{3/5}{-4/5} = -\frac{3}{4}$.

Example

If you're given the value of $\cos\theta$ and the quadrant of the angle, show how you can find $\sin\theta$ and $\tan\theta$

Given the value of $\cos\theta$, we can find $\sin\theta$ using the Pythagorean identity, $\sin^2(\theta) + \cos^2(\theta) = 1$. Solving for $\sin\theta$, we get $\sin^2\theta = 1 - \cos^2\theta$, and $\sin\theta = \pm\sqrt{1-\cos^2\theta}$. For instance, if $\cos\theta = \frac{12}{13}$, then $\sin\theta = \pm\sqrt{1-\left(\frac{12}{13}\right)^2} = \pm\sqrt{1-\frac{144}{169}} = \pm\sqrt{\frac{25}{169}} = \pm\frac{5}{13}$. To decide whether the sine is positive or negative, we take into account the quadrant of the angle: if the angle is in quadrant 1 or 2, $\sin\theta$ is positive, and if the angle is in quadrant 3 or 4, $\sin\theta$ is negative.

Since the tangent is equal to the sine divided by the cosine, once we know the sine and the cosine finding the tangent is simple. Again, if $\cos\theta = \frac{12}{13}$, and if the angle is in quadrant 1 so the sine is positive, then $\tan\theta = \frac{5/13}{12/13} = \frac{5}{12}$.

Example

If you're given the value of $\tan\theta$ and the quadrant of the angle, show how you can find $\sin\theta$ and $\cos\theta$

Given the value of $\tan\theta$, one way we can find $\sin\theta$ and $\cos\theta$ is by using the identity $1+\tan^2\theta=\sec^2\theta$. So $\sec\theta=\pm\sqrt{1+\tan^2\theta}$, and since $\cos\theta=1/\sec\theta$, $\cos\theta=\pm(1+\tan^2\theta)^{-1/2}$. For instance, if $\tan\theta=\frac{15}{8}$, then $\cos\theta=\pm\left(1+\left(\frac{15}{8}\right)^2\right)^{-1/2}=\pm\left(1+\frac{225}{64}\right)^{-1/2}=\pm\left(\frac{289}{64}\right)^{-1/2}=\pm\sqrt{\frac{64}{289}}=\pm\frac{8}{17}$. To decide whether the cosine is positive or negative, we take into account the quadrant of the angle: if the angle is in quadrant 1 or 4, $\cos\theta$ is positive, and if the angle is in quadrant 2 or 3, $\cos\theta$ is negative.

Since the tangent is equal to the sine divided by the cosine, once we know the cosine and the tangent finding the sine is simple: we just multiply the cosine by the tangent. Again, if $\tan\theta=\frac{15}{8}$, and if the angle is in quadrant 1 so the cosine is positive, then $\sin\theta=\left(\frac{15}{8}\right)\left(\frac{8}{17}\right)=\frac{15}{17}$.

Deriving the Addition Formula for the Sine Function

The addition formula for the sine function is $\sin(a+b)=\sin a\cos b+\sin b\cos a$. There are many different ways this can be proven. One proof (at least for a and b between 0 and $\frac{\pi}{2}$) uses the following diagram:

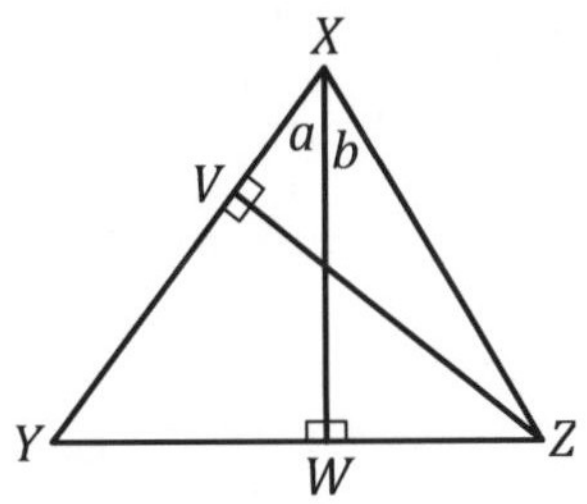

Consider first right triangle XWZ. By the right-triangle definition of the sine and cosine, $WZ=XZ\sin b$ and $WX=XZ\cos b$. Similarly, considering right triangle XWY, we find that $WY=XY\sin a$ and $WX=XY\cos a$. Since the area of a triangle is one half the base times the height, the area of triangle XWZ is $\frac{1}{2}WZ\cdot WX=\frac{1}{2}(XZ\sin b)(XY\cos a)$, and the area of triangle XWY is $\frac{1}{2}WY\cdot WX=\frac{1}{2}(XY\sin a)(XZ\cos b)$. Now, from right triangle XVZ, we see that $VZ=XZ\sin(a+b)$. If we take XY to be the base of triangle XYZ, we therefore find that the area of triangle XYZ is equal to $\frac{1}{2}XY\cdot VZ=\frac{1}{2}XY(XZ\sin(a+b))$. Since the area of XYZ must be the sum of the areas of XWY and XWZ, we find $\frac{1}{2}XY(XZ\sin(a+b))=\frac{1}{2}(XY\sin a)(XZ\cos b)+\frac{1}{2}(XZ\sin b)(XY\cos a)$, or, canceling $\frac{1}{2}XY\cdot XZ$ from each term, $\sin(a+b)=\sin a\cos b+\sin b\cos a$.

DERIVING THE ADDITION FORMULA FOR THE COSINE FUNCTION

The addition formula for the cosine function is $\cos(a + b) = \cos a \cos b - \sin a \sin b$. There are many different ways this can be proven; one uses the unit circle, as follows:

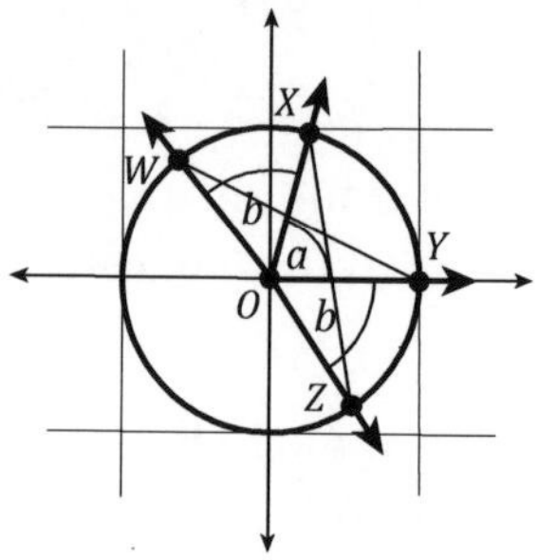

Note that the angles between OX and OZ and between OW and OY are both equal to $a + b$, so $XZ = WY$. By the distance formula, $XZ^2 = (\cos a - \cos b)^2 + (\sin a + \sin b)^2 = \cos^2 a + \cos^2 b - 2\cos a \cos b + \sin^2 a + \sin^2 b + 2\sin a \sin b = (\sin^2 a + \cos^2 a) + (\sin^2 b + \cos^2 b) + 2(\sin a \sin b - \cos a \cos b) = 2 + 2(\sin a \sin b - \cos a \cos b)$. Similarly, $WY^2 = (1 - \cos(a + b))^2 + (\sin(a + b))^2 = 1 + \cos^2(a + b) - 2\cos(a + b) + \sin^2(a + b) = 1 + (\sin^2(a + b) + \cos^2(a + b)) - 2\cos(a + b) = 2 - 2\cos(a + b)$. Setting these equal to each other, we get $2 + 2(\sin a \sin b - \cos a \cos b) = 2 - 2\cos(a + b)$, so $\cos(a + b) = \cos a \cos b - \sin a \sin b$.

DERIVING THE ADDITION FORMULA FOR THE TANGENT FUNCTION

The addition formula for the tangent function is $\tan(a + b) = \frac{\tan a + \tan b}{1 - \tan a \tan b}$. This can be proven by taking the ratio of the addition formulas for the sine and cosine formulas: $\tan(a + b) = \frac{\sin(a+b)}{\cos(a+b)} = \frac{\sin a \cos b + \cos a \sin b}{\cos a \cos b - \sin a \sin b}$. If we divide both the numerator and the denominator by $\cos a \cos b$, the numerator becomes $\frac{\sin a \cos b}{\cos a \cos b} + \frac{\cos a \sin b}{\cos a \cos b} = \frac{\sin a}{\cos a} + \frac{\sin b}{\cos b} = \tan a + \tan b$, and the denominator becomes $\frac{\cos a \cos b}{\cos a \cos b} - \frac{\sin a \sin b}{\cos a \cos b} = 1 - \tan a \tan b$, giving a ratio of $\frac{\tan a + \tan b}{1 - \tan a \tan b}$. (There are ways of proving the addition formula for the tangent function directly without relying on the formulas for the sine and cosine, but they're substantially more complicated.)

Example

Determine how you could use trigonometric identities to solve a problem such as finding the sine of θ in the diagram below

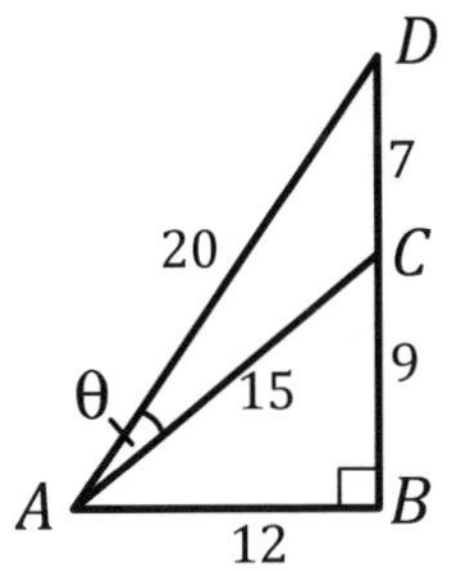

One way to find $\sin\theta$ in the given diagram is by using the sine subtraction formula. We may not be able to find $\sin\theta$ directly, since it's not part of a right triangle, but we can find the sines of angles DAB and CAB, which *are* angles in right triangles. Using the fact that the sine of an angle in a right triangle is equal to the length of the opposite leg divided by the length of the hypotenuse, we find $\sin(\angle DAB) = (7 + 9)/20 = 4/5$, and $\sin(\angle CAB) = 9/15 = 3/5$. Likewise, by dividing the length of the *adjacent* leg by the length of the hypotenuse, we find $\cos(\angle DAB) = 12/20 = 3/5$, and $\cos(\angle CAB) = 12/15 = 4/5$. So, using the subtraction formula for the sine, $\sin(\alpha - \beta) = \sin\alpha\cos\beta - \cos\alpha\sin\beta$, we have $\sin\theta = \sin(\angle DAB - \angle CAB) = \sin(\angle DAB)\cos(\angle CAB) - \cos(\angle DAB)\sin(\angle CAB) = (4/5)(4/5) - (3/5)(3/5) = (16/25) - (9/25) = 7/25$.

EVALUATING A TRIGONOMETRIC FUNCTION SUCH AS $\cos 15°$ OR $\tan 7\pi/12$

Although $15°$ and $7\pi/12$ are not angles the trigonometric functions of which are commonly memorized, they can be written as sums or differences of angles with trigonometric functions that *are* well known: $15° = 45° - 30°$, and $7\pi/12 = \pi/3 + \pi/4$. (These sums or differences are not unique; we could also have used, for instance, $15° = 60° - 45°$ or $7\pi/12 = 3\pi/4 - \pi/6$.) We can therefore use the sum and difference formulas to solve for the trigonometric functions of these angles.

The cosine difference formula is $\cos(a - b) = \cos a\cos b + \sin a\sin b$. In this case, $\cos(45° - 30°) = (\cos 45°)(\cos 30°) + (\sin 45°)(\sin 30°) = \left(\frac{\sqrt{2}}{2}\right)\left(\frac{1}{2}\right) + \left(\frac{\sqrt{2}}{2}\right)\left(\frac{\sqrt{3}}{2}\right) = (\sqrt{2} + \sqrt{6})/4$.
The tangent sum formula is $\tan(a + b) = \frac{\tan a + \tan b}{1 - \tan a\tan b}$. In this case, $\tan(\pi/3 + \pi/4) = \frac{\tan(\pi/3) + \tan(\pi/4)}{1 - (\tan\pi/3)(\tan\pi/4)} = \frac{\sqrt{3} + 1}{1 - (\sqrt{3})(1)} = \frac{1 + \sqrt{3}}{1 - \sqrt{3}}$.

Practice Test #1

Practice Questions

1. The function f is defined as $f(x) = \sqrt{x-3}$. Its domain is $x \geq 3$, and its range is $f(x) \geq 0$. Which of the following is true of f?

a. If $x \geq 3$, then $f(x) \geq 0$.
b. It assigns exactly one value to every positive value of x.
c. The range of the function is $f(3)$.
d. The value of $f(3)$ is undefined.

2. If $g(x) = 3x + x + 5$, evaluate $g(2)$.

a. $g(2) = 8$
b. $g(2) = 9$
c. $g(2) = 13$
d. $g(2) = 17$

3. The function $S(r) = 4\pi r^2$ gives the surface area of a sphere of radius r. What is the surface area of a sphere of radius 4?

a. 8π
b. 16π
c. 32π
d. 64π

4. Calculate the fifth term of the sequence defined as $f(0) = 2$, $f(n + 1) = 2f(n) - 1$ for $n \geq 1$.

a. 9
b. 17
c. 32
d. 33

5. A metal ball is thrown up in the air. Its height above the ground (in feet) t seconds after it is thrown is given by the function $h(t)$. The equation $y = h(t)$ is graphed below.

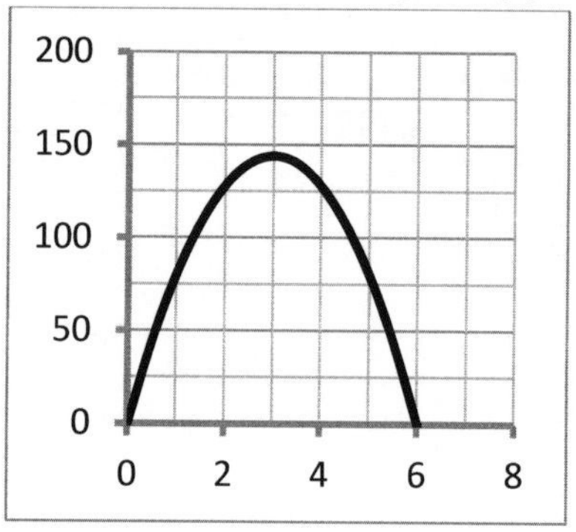

How many seconds after it is thrown does the metal ball reach its highest point?

a. 3 seconds
b. 6 seconds
c. 50 seconds
d. 144 seconds

6. Which of the following would be an appropriate domain for the function $t(n)$, the average temperature (in °F) during the nth month of the year in Washington, DC?

a. $0 \leq n \leq 12$
b. $1 \leq n \leq 12$
c. $10 \leq n \leq 90$
d. $40 \leq n \leq 60$

7. Calculate the average rate of change of the function $f(x) = -3x + 1$, over the range from $x = 1$ to $x = 5$.

a. −4
b. −3
c. 3
d. 4

8. Graph the function $f(x) = 2x - 4$. Show its x- and y-intercepts.

a.
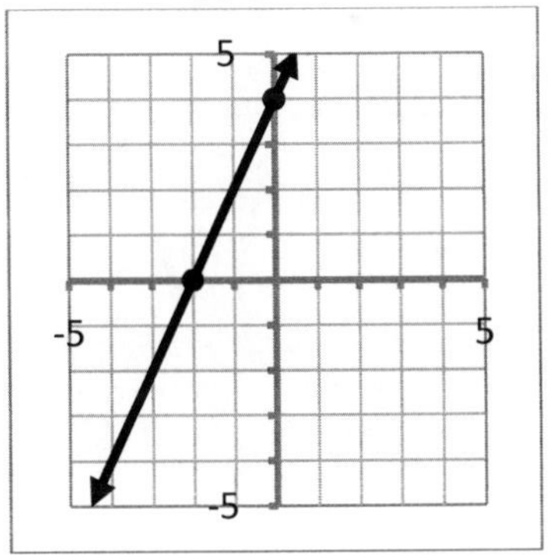

b.
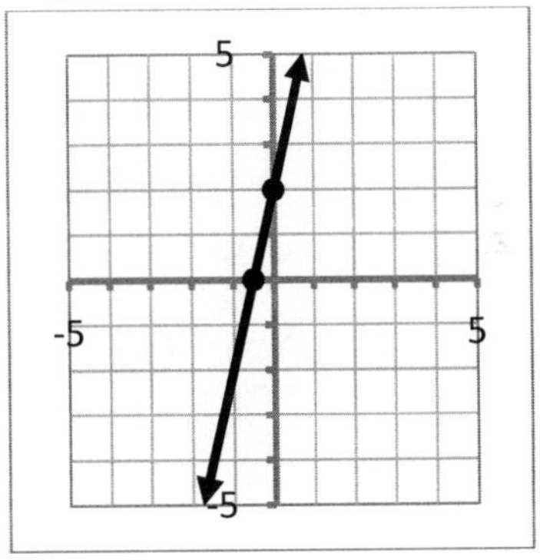

c.
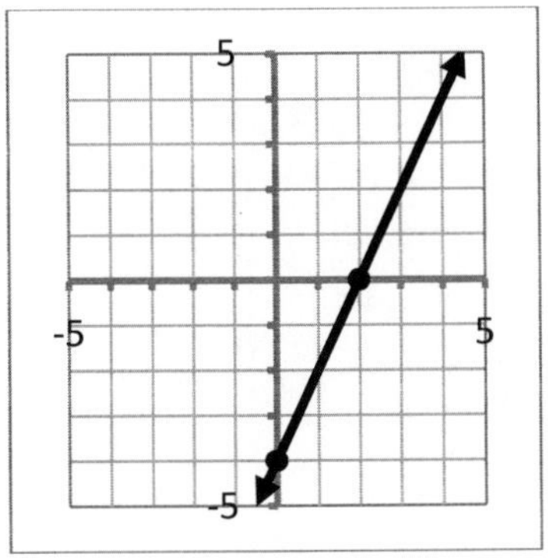

d.
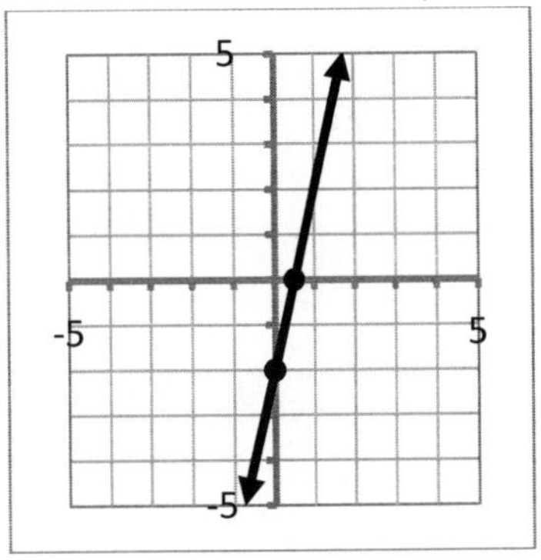

9. **Graph the function $g(x) = x^2 - 2x - 3$. Show its minimum.**

a.

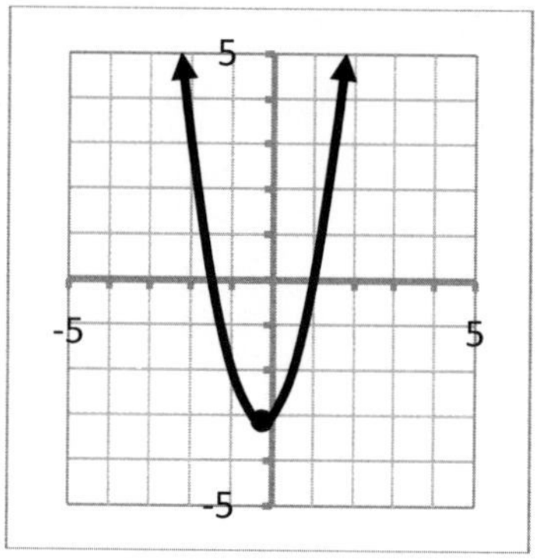

b.

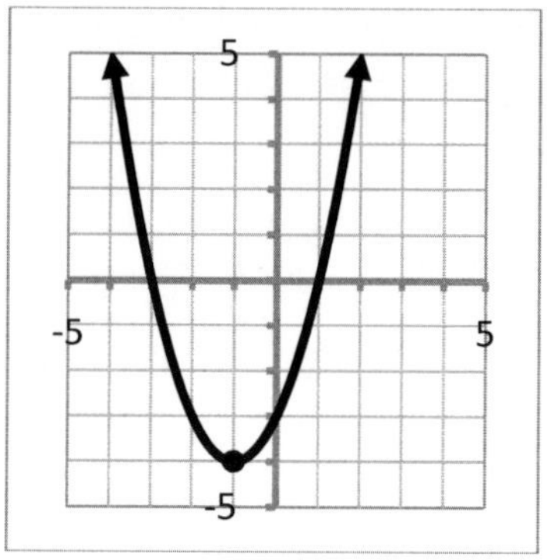

c.

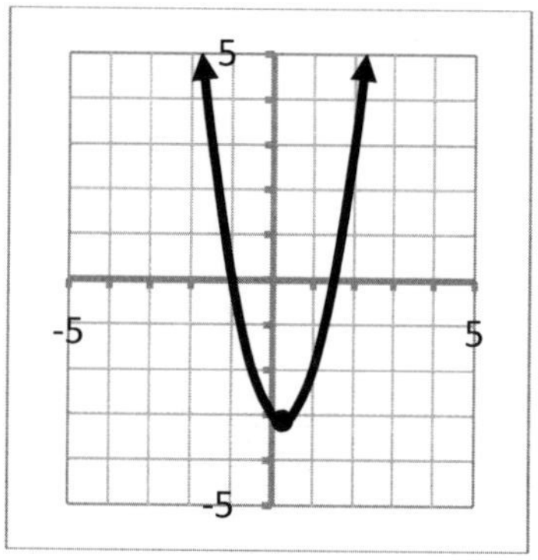

d.

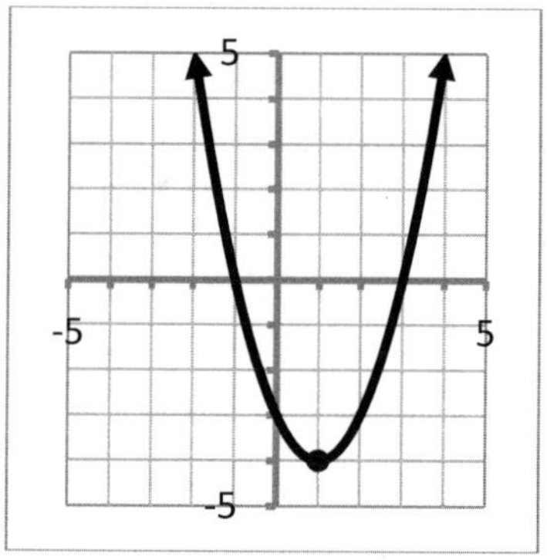

10. **Graph the piecewise-defined function $h(x) = \begin{cases} \frac{1}{2}x, & x < 2 \\ 4 - x, & x \geq 2 \end{cases}$.**

a.

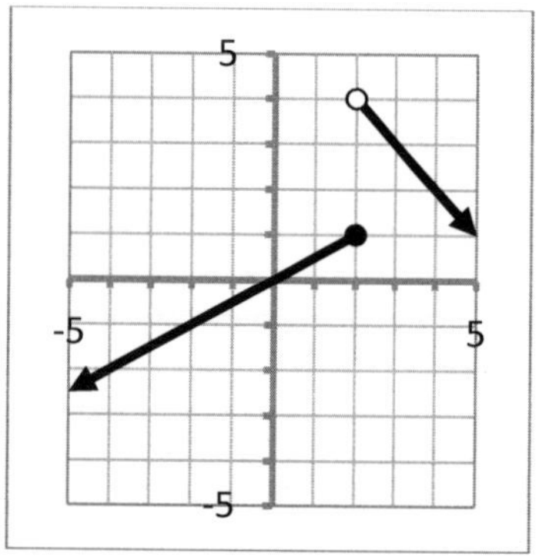

b.

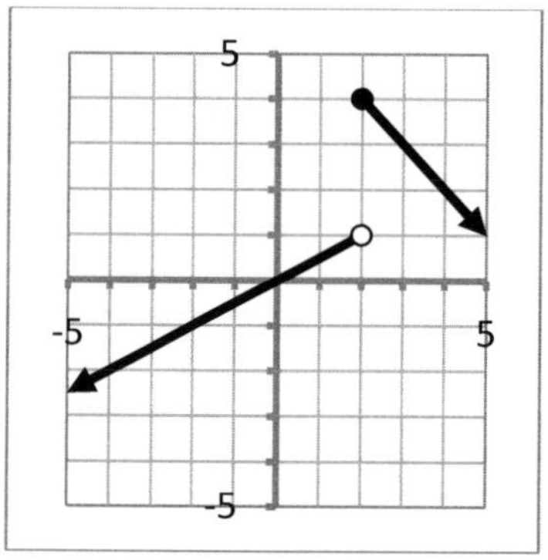

c.

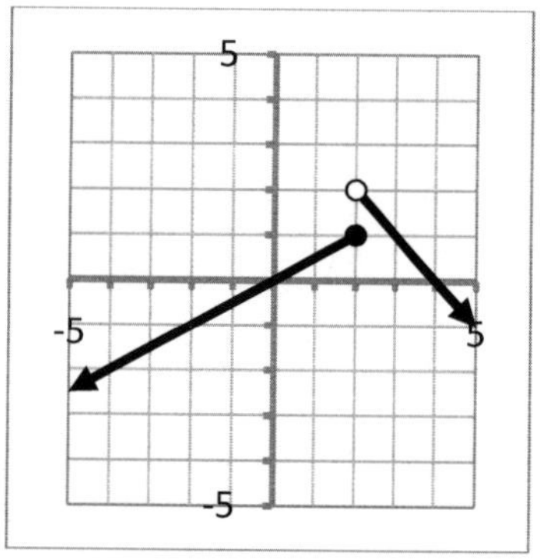

d.

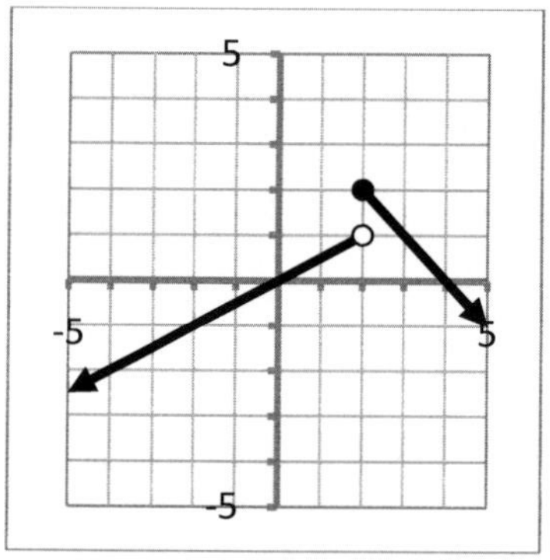

11. **Graph the function $f(x) = x^3 + x^2 - 4x - 4$. Show the function's zeroes.**

a.

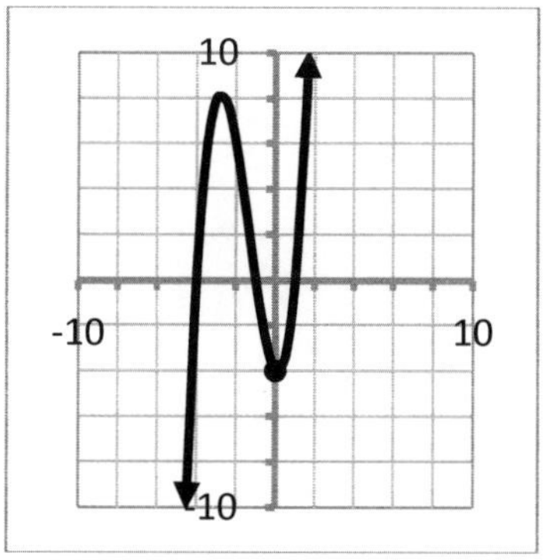

b.

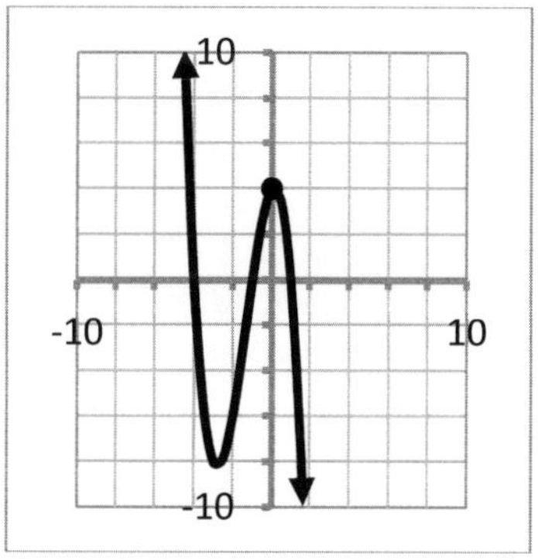

c.

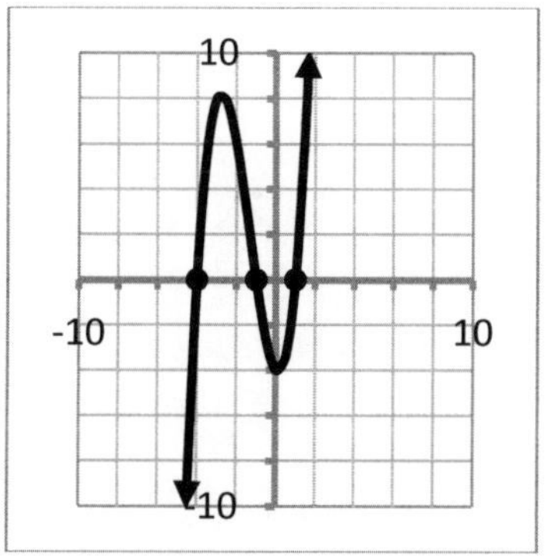

d.

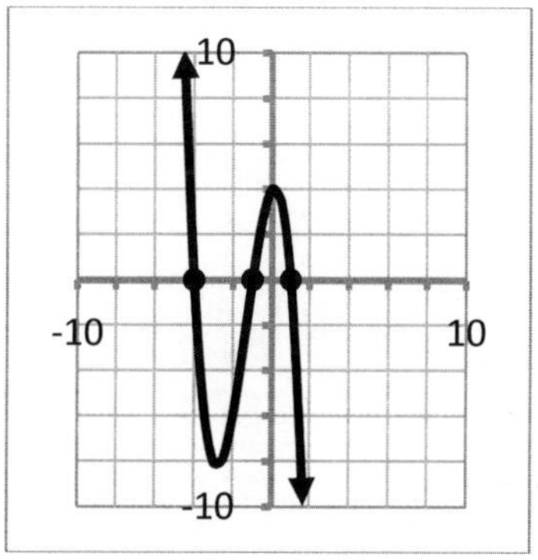

12. **Graph the function $g(x) = \frac{3}{2x-4}$. Show its zeroes and asymptotes if there are any.**

a.

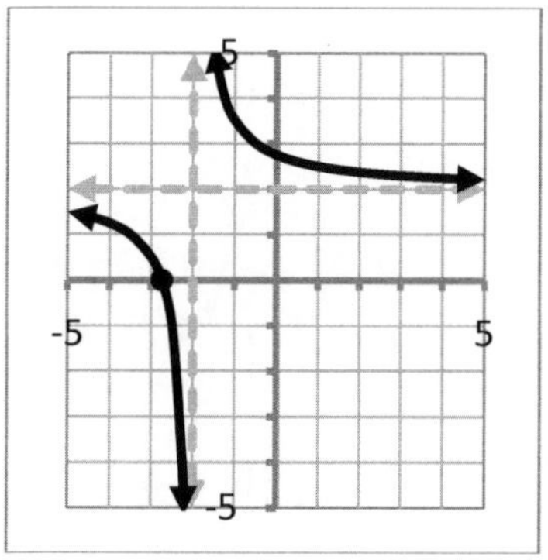

b.

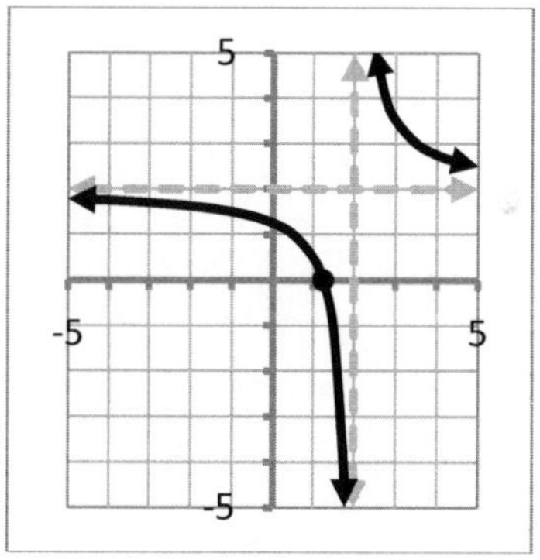

c.

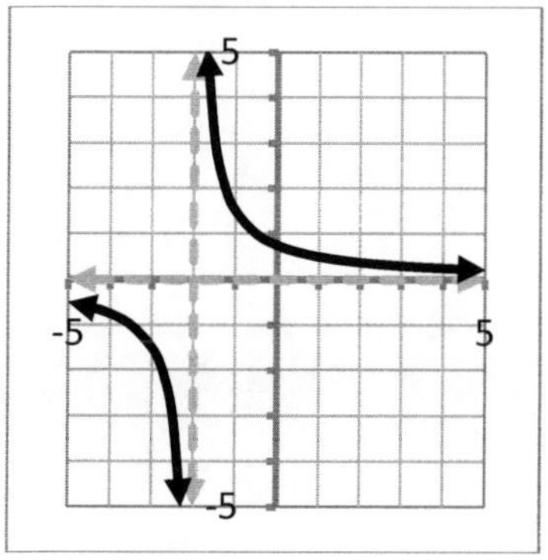

d.

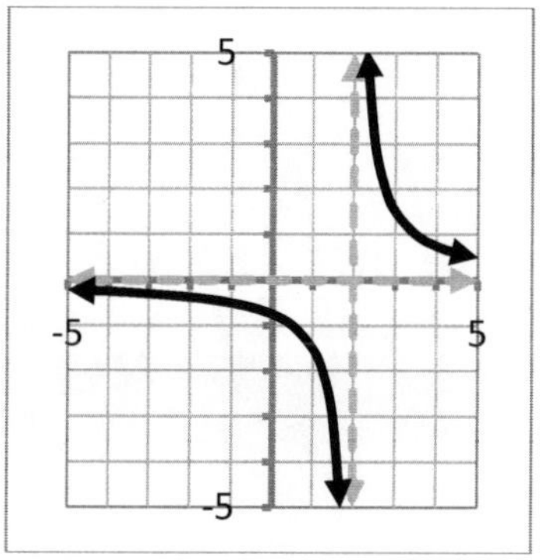

13. Graph the function $h(x) = \log_2 2(x + 3)$. Show its x- and y-intercepts if there are any.

a. 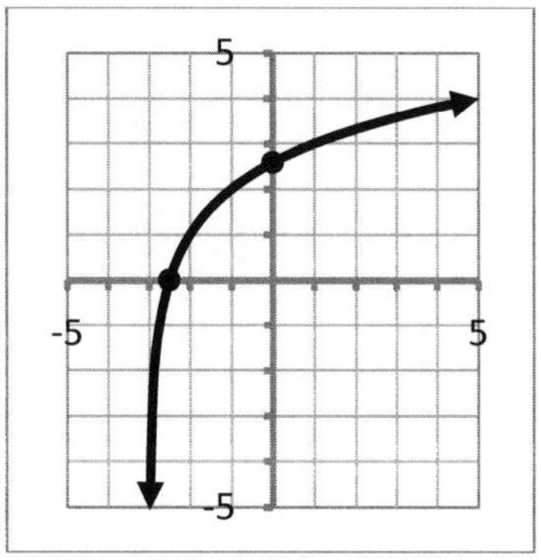

b. 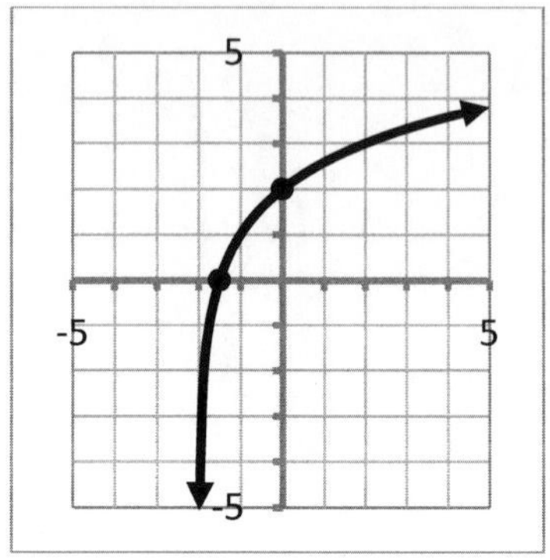

c. 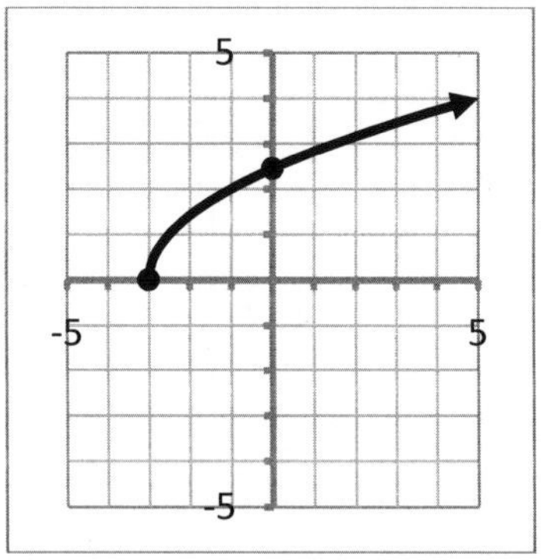

d. 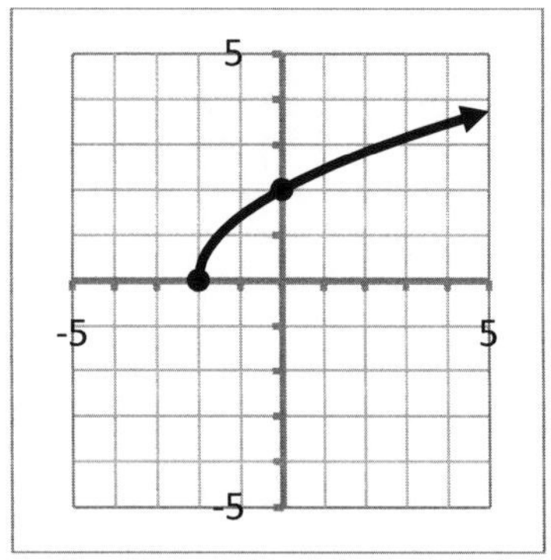

14. Graph the function $f(x) = 2\sin(3x - \frac{\pi}{2})$.

a. 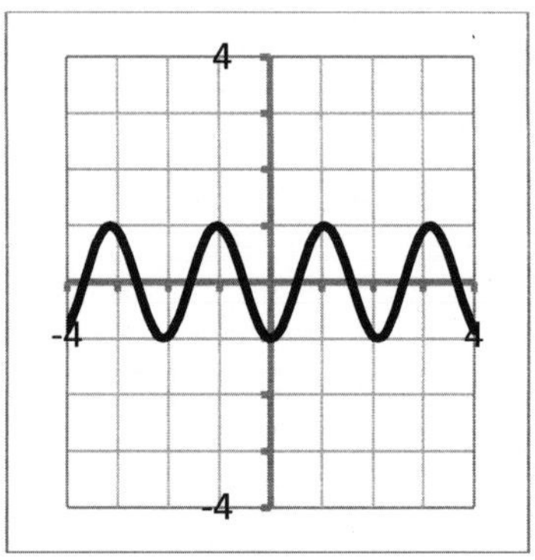

b. 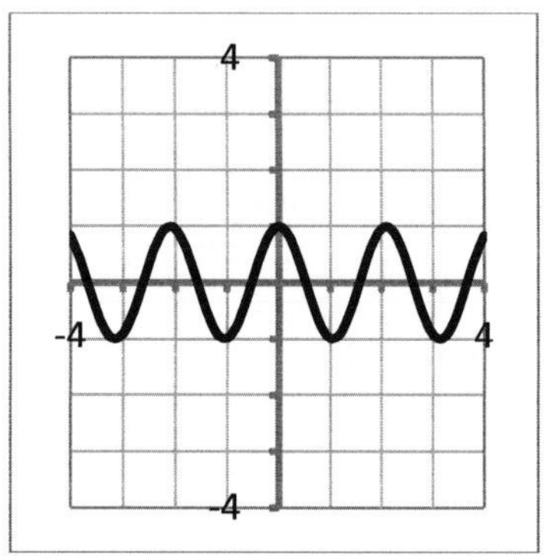

c. 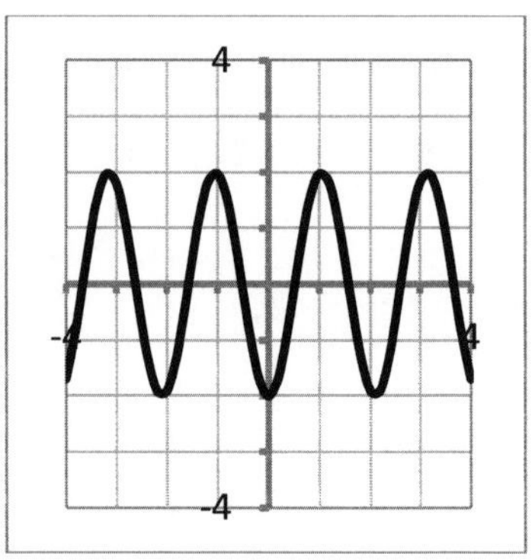

d. 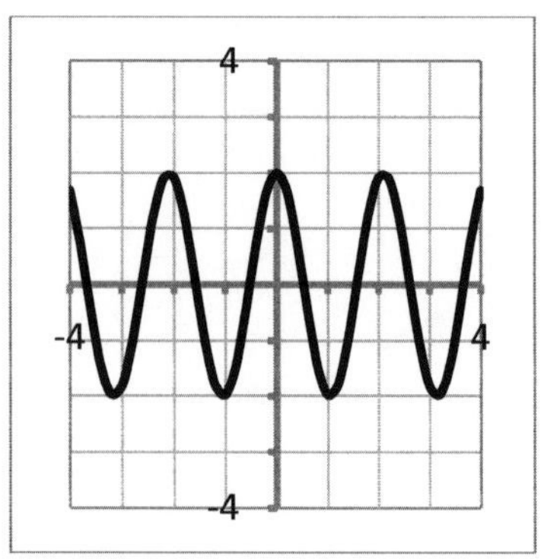

15. Use factoring to identify the zeroes of the function $f(x) = x^2 + 5x - 24$.

a. -24
b. -8 and 3
c. -6 and 4
d. 24

16. The dollar-value V of a particular interest-earning investment after t years is modeled by the exponential function $V = 200(1.04)^t$. What is the yearly interest of the investment?

a. 2%
b. 4%
c. 104%
d. 200%

17. Which function has the same x-intercept as the function graphed below?

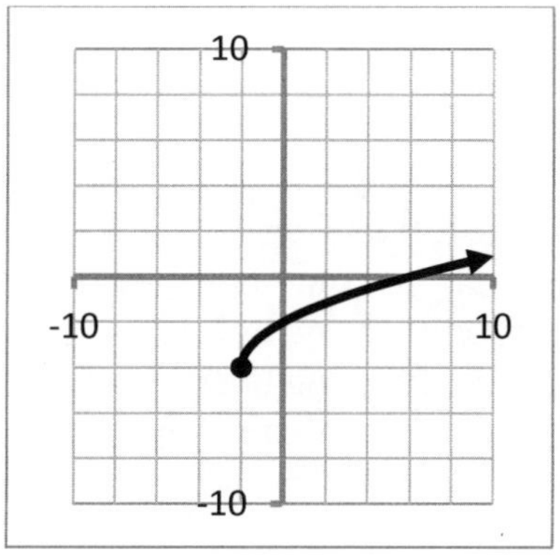

a. $g(x) = -4x - 12$
b. $g(x) = x^2 - 12x + 36$
c. $g(x) = x^3 + 6x^2 - 2x - 2$
d. $g(x) = \frac{6}{x-3}$

18. A theater will sell 500 tickets to a play if it charges $10 per ticket. Furthermore, every time it raises the price by one dollar, it will sell 50 fewer tickets because some people will think it too expensive. Which of the following functions $t(d)$ represents the number of tickets the theater will sell if it charges d dollars per ticket?

a. $t(d) = -50d + 10$
b. $t(d) = -50d + 1000$
c. $t(d) = 50d$
d. $t(d) = 50d + 10$

19. A taxi ride costs $4.25 for the first mile and $0.70 for each mile after the first. Which of the following functions $c(d)$ gives the total cost (in dollars) of traveling d miles (assuming that $d \geq 1$)?

a. $c(d) = 3.55 + 0.70d$
b. $c(d) = 3.55 + 0.70(d - 1)$
c. $c(d) = 4.25 + 0.70d$
d. $c(d) = 4.25 + 0.70(d - 1)$

20. A pump fills a cylindrical tank with water at a constant rate. The function $L(g) = 0.3g$ represents the water level of the tank (in feet) after g gallons are pumped into the tank. The function $w(t) = 1.2t$ represents the number of gallons that can be pumped into the tank in t minutes. Write a function $L(t)$ for the water level of the tank after t minutes.

a. $L(t) = 0.25t$
b. $L(t) = 0.36t$
c. $L(t) = 0.9t$
d. $L(t) = 3.6t$

21. Grains of wheat are placed on the squares of a chessboard so that one grain is placed on the first square, two on the second, four are on the third, and so on, continuing so that twice as many grains are placed on each subsequent square. Write a formula for the number of wheat grains that are placed on the *n*th square.

a. $g(n) = 2(n - 1)$
b. $g(n) = 2n$
c. $g(n) = n^2$
d. $g(n) = 2^{n-1}$

22. Compare the graphs of $f(x) = 3^x$ and $g(x) = 3^{x+1}$.

a. The graph of g is the graph of f shifted one unit down.
b. The graph of g is the graph of f shifted one unit up.
c. The graph of g is the graph of f shifted one unit to the left.
d. The graph of g is the graph of f shifted one unit to the right.

23. Compare the graphs of $f(x) = \sqrt{x}$ and $g(x) = 5\sqrt{x}$.

a. The graph of g is the graph of f shifted five units to the right.
b. The graph of g is the graph of f shifted five units up.
c. The graph of g is the graph of f stretched horizontally
d. The graph of g is the graph of f stretched vertically.

24. What is the inverse of the function $f(x) = \frac{3-x}{2}$?

a. $f^{-1}(x) = \frac{1}{2}x + 3$
b. $f^{-1}(x) = 2x + 3$
c. $f^{-1}(x) = 3 - \frac{1}{2}x$
d. $f^{-1}(x) = 3 - 2x$

25. What is the inverse of the function $f(x) = x^3 + 8$?

a. $f^{-1}(x) = \sqrt[3]{x - 2}$
b. $f^{-1}(x) = \sqrt[3]{x} - 2$
c. $f^{-1}(x) = \sqrt[3]{x - 8}$
d. $f^{-1}(x) = \sqrt[3]{x} - 8$

26. The inverse trigonometric function $\cos^{-1} x$ is defined as the inverse of $\cos x$. What is the composition of $\cos^{-1} t$ and $\cos t$?

a. $\cos^{-1}(\cos t) = 0$
b. $\cos^{-1}(\cos t) = 1$
c. $\cos^{-1}(\cos t) = t$
d. $\cos^{-1}(\cos t) = x$

27. The equation $y = g(x)$ is graphed below.

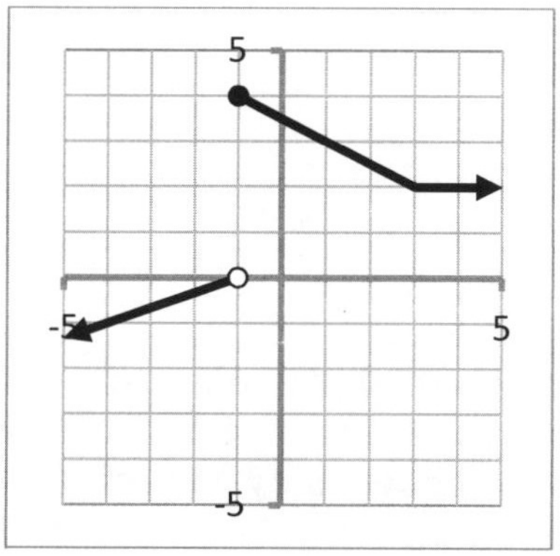

What is the value of $g^{-1}(3)$?

a. $g^{-1}(3) = -1$
b. $g^{-1}(3) = 1$
c. $g^{-1}(3) = 2$
d. $g^{-1}(3) = 3$

28. The table below displays the value of $h(x)$ for nine different values of x.

x	–4	–3	–2	–1	0	1	2	3	4
h(x)	–16	–8	–4	–2	–1	–1	0	2	6

What is the value of $h^{-1}(-2)$?

a. $h^{-1}(-2) = -4$
b. $h^{-1}(-2) = -1$
c. $h^{-1}(-2) = 0$
d. $h^{-1}(-2) = 3$

29. The function $f(x) = x^2 - 4x + 3$ is not invertible if the domain is all real numbers. Which domain makes the function invertible?

a. $x \geq -1$
b. $x \geq 2$
c. $x \leq 3$
d. $x \leq 4$

30. Solve the equation $\log_2(x^2 - 8) = 3$ for x.

a. $x = 0$
b. $x = \pm 4$
c. $x = \pm 8$
d. $x = 16$

31. Exponential functions grow by equal factors over equal intervals. By what factor does the exponential function $f(x) = 3 \cdot 2^x$ grow by over every interval whose length is 3?

a. By a factor of 6
b. By a factor of 8
c. By a factor of 18
d. By a factor of 24

32. A linear function can be used to convert a temperature from Fahrenheit to Celsius. For example, you can use it to convert 32°F to 0°C and 68°F to 20°C. Use this information to convert 104°F to Celsius.

a. 25°C
b. 30°C
c. 40°C
d. 50°C

33. The growth of bacteria can be modeled with an exponential function. If you start with 160 bacteria in a Petri dish, in two days, there will be 360 bacteria. And in three days, there will be 540 bacteria. If this continues, how many bacteria will there be in the Petri dish in five days?

a. 740 bacteria
b. 810 bacteria
c. 900 bacteria
d. 1215 bacteria

34. The table below displays the value of the linear function $g(x)$ for different values of x.

x	−2	−1	0	1	2
$g(x)$	10	7	4	1	−2

Write an explicit formula for $g(x)$.

a. $g(x) = -4x + 3$
b. $g(x) = -4x + 4$
c. $g(x) = -3x + 3$
d. $g(x) = -3x + 4$

35. The exponential function $h(x)$ is graphed below.

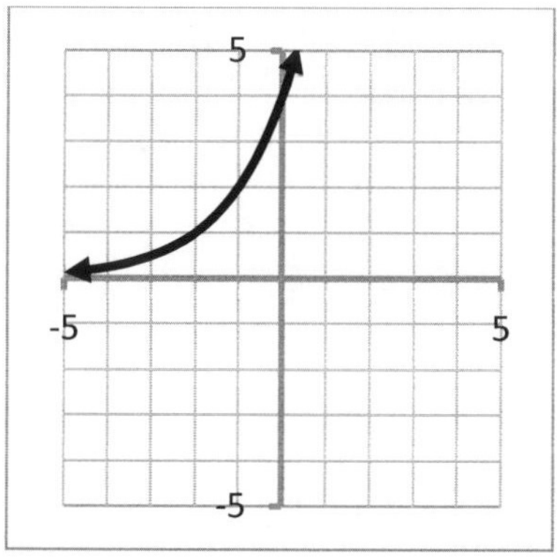

Write an explicit formula for $h(x)$.

a. $h(x) = 2^x$
b. $h(x) = 4 \cdot 2^x$
c. $h(x) = 4^x$
d. $h(x) = 2 \cdot 4^x$

36. Four functions are graphed below.

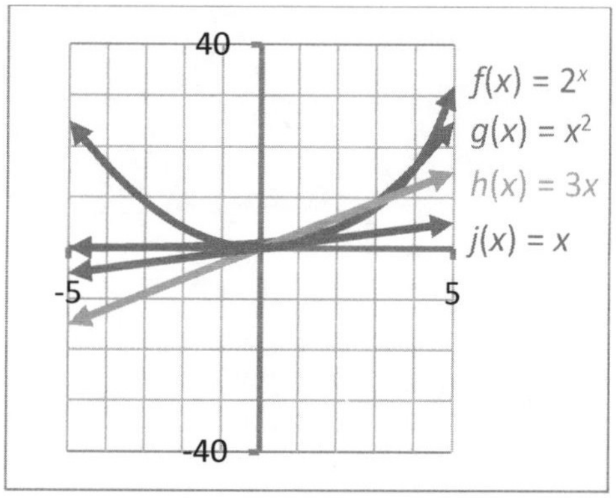

Which function eventually exceeds the other three as x increases?

a. $f(x) = 2^x$
b. $g(x) = x^2$
c. $h(x) = 3x$
d. $j(x) = x$

37. Solve the exponential equation $3e^{2t} = 12$ for t.

a. $t = \frac{\ln 4}{2}$
b. $t = \frac{\ln 12}{6}$
c. $t = \ln 2$
d. $t = 2\ln 4$

38. A librarian makes time-and-a-half for each hour that he works over 40 hours per week. The linear function $s(h) = 27h + 720$ represents his weekly salary (in dollars) if he works h hours more than 40 hours that week. For example, his weekly salary is $s(10) = 990$ dollars if he works 50 hours in one week (because 50 hours means 10 hours of overtime). What is his hourly salary after he has already worked the initial 40 hours in one week?

a. $7.20 per hour
b. $18.00 per hour
c. $27.00 per hour
d. $72.00 per hour

39. John puts his money into a bank account that pays monthly interest. His monthly balance (in dollars) after t months is given by the exponential function $b(t) = 315(1.05)^t$. How much money did John initially put into the account?

a. $5
b. $105
c. $300
d. $315

40. In the figure below, circle O is a unit circle, and the measure of $\angle AOB$ is $\frac{\pi}{3}$.

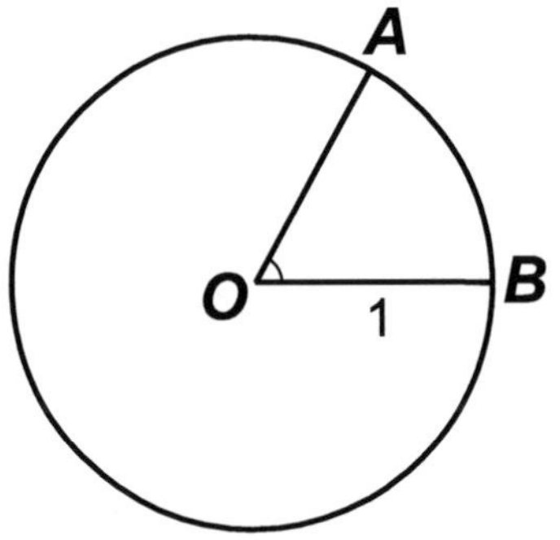

What is the length of $\widehat{AB}$?

a. $\frac{\pi}{6}$
b. $\frac{\pi}{3}$
c. $\frac{2\pi}{3}$
d. π

41. A unit circle and an angle are graphed on the coordinate plane below.

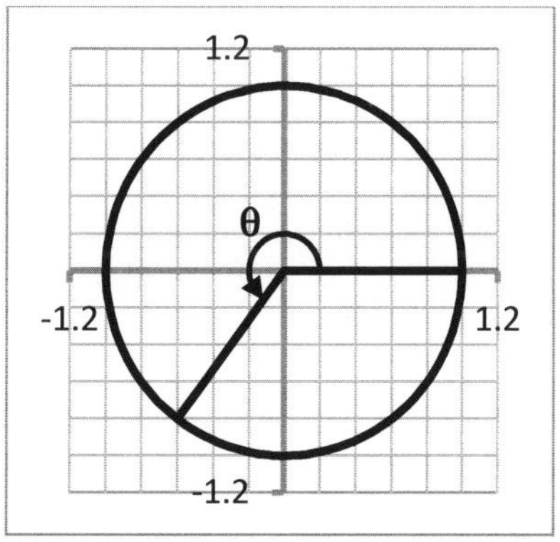

Use the graph to calculate the value of sin θ.

a. $\sin\theta = -0.8$
b. $\sin\theta = -0.6$
c. $\sin\theta = 0.6$
d. $\sin\theta = 0.8$

42. Calculate the value of $\tan\frac{\pi}{6}$.

a. $\tan\frac{\pi}{6} = \frac{1}{2}$
b. $\tan\frac{\pi}{6} = \frac{\sqrt{3}}{3}$
c. $\tan\frac{\pi}{6} = \sqrt{3}$
d. $\tan\frac{\pi}{6} = 2$

43. Express $\sin(\pi - x)$ as a function with an argument of x.

a. $\sin(\pi - x) = -\cos x$
b. $\sin(\pi - x) = -\sin x$
c. $\sin(\pi - x) = \cos x$
d. $\sin(\pi - x) = \sin x$

44. Which of the following describes the properties of cos x?

a. It is aperiodic and has even symmetry.
b. It is aperiodic and has odd symmetry.
c. It is periodic and has even symmetry.
d. It is periodic and has odd symmetry.

45. The height of a weight suspended from a spring can be modeled with a cosine function. The weight is initially at its lowest point of 5 inches, and it takes 0.5 seconds for the weight to go from that point to its highest height of 25 inches, and then it takes 0.5 seconds to go back again. Write a function $h(t)$ that represents the height of the weight (in inches) after t seconds.

a. $h(t) = 10\cos(2t) + 5$
b. $h(t) = -10\cos(2t) + 15$
c. $h(t) = -10\cos(2\pi t) + 5$
d. $h(t) = -10\cos(2\pi t) + 15$

46. The function *f*(*x*) = cos *x* is not invertible if the domain is all real numbers. Which domain makes the function invertible?

a. $0 \leq x \leq \pi$
b. $0 \leq x \leq 2\pi$
c. $\frac{\pi}{2} \leq x \leq \frac{3\pi}{2}$
d. $\frac{\pi}{2} \leq x \leq \frac{5\pi}{2}$

47. John is sitting in a car on the London Eye, the tallest Ferris wheel in Europe. The trigonometric function $h(t) = -60\cos\left(\frac{\pi t}{15}\right) + 70$ represents his height above the ground (in meters) *t* minutes after he starts moving. Use an inverse function to determine when he is 90 m above the ground for the first time. Round your answer to the nearest minute.

a. After about 6 minutes
b. After about 9 minutes
c. After about 18 minutes
d. After about 29 minutes

48. If $\sin\theta = \frac{4}{5}$ and θ is in quadrant I, then what is tan θ?

a. $\tan\theta = \frac{9}{16}$
b. $\tan\theta = \frac{3}{5}$
c. $\tan\theta = \frac{4}{3}$
d. $\tan\theta = \frac{16}{9}$

49. Calculate sin 75° using the addition formula for sine.

a. $\sin 75° = -\frac{1}{2}$
b. $\sin 75° = \frac{\sqrt{2}-\sqrt{6}}{4}$
c. $\sin 75° = \frac{\sqrt{2}}{2}$
d. $\sin 75° = \frac{\sqrt{2}+\sqrt{6}}{4}$

50. If $\tan x = \frac{4}{3}$ and $\tan y = \frac{1}{2}$, then what is tan (*x* – *y*)?

a. $\tan(x-y) = \frac{1}{2}$
b. $\tan(x-y) = \frac{11}{10}$
c. $\tan(x-y) = \frac{5}{2}$
d. $\tan(x-y) = \frac{11}{2}$

Answers and Explanations

Math.Content.F.IF.1

1. A: The function f assigns to each element of the domain exactly one element of the range. Consequently, if x is in the domain $x \geq 3$, then the value of f is in the range $f(x) \geq 0$. Therefore, the correct answer is choice A. On the other hand, choice B is incorrect because f is not defined for some positive values of x, like $x = 1$. Choice C is incorrect because the range of f is $f(x) \geq 0$, not $f(3)$. Finally, choice D is incorrect because the value of $f(3)$ is defined since $x = 3$ is in the domain of f.

Math.Content.F.IF.2

2. C: To evaluate $g(2)$, substitute 2 for every occurrence of x in the equation $g(x) = 3x + x + 5$. Then simplify the result using order of operations.

$$g(2) = 3(2) + (2) + 5$$
$$g(2) = 6 + 2 + 5$$
$$g(2) = 13$$

Math.Content.F.IF.2

3. D: The surface area will be given by the expression $S(4)$. To calculate this value, substitute 4 for r in the equation $S(r) = 4\pi r^2$. Then simplify the result using order of operations:

$$\begin{aligned} S(4) &= 4\pi(4)^2 \\ &= 4\pi \cdot 16 \\ &= 64\pi \end{aligned}$$

Therefore, the surface area of the sphere is 64π.

Math.Content.F.IF.3

4. B: Since the function is defined recursively, you need to calculate all of the first five terms. The first term is already given in the problem as $f(0) = 2$. To calculate the second term, $f(1)$, which we can write as $f(0 + 1)$ in order to use the recursive definition, substitute 0 for n into the equation $f(n + 1) = 2f(n) - 1$:

$$f(1) = f(0 + 1) = 2f(0) - 1$$
$$f(1) = 2f(0) - 1$$

Since $f(0) = 2$, substitute 2 in for $f(0)$ and simplify the result:

$$\begin{aligned} f(1) &= 2(2) - 1 \\ &= 3 \end{aligned}$$

Thus, the second term is 3. To calculate the third term, $f(2)$, write this as $f(1 + 1)$ and substitute 1 for n and 3 for $f(1)$ into the same equation $f(n + 1) = 2 \cdot f(n) - 1$ and simplify the result:

$$\begin{aligned} f(1 + 1) &= 2f(1) - 1 \\ f(2) &= 2(3) - 1 \\ &= 5 \end{aligned}$$

Continuing in this way, you will find that the first five terms of the sequence are 2, 3, 5, 9, 17. Therefore, the fifth term of the sequence is 17.

Math.Content.F.IF.4

5. A: The metal ball reaches its highest point when h is at its maximum. From the graph, you can tell that the maximum is about 144. Since $h(t) = 144$ when $t = 3$, the ball reaches its highest point 3 seconds after it is thrown.

Math.Content.F.IF.5

6. B: The domain of a function is the set of all possible input values that can be evaluated by a function. In the given function, the input is n, which represents the month of the year (i.e. $n = 1$ represents January, $n = 2$ represents February, etc.). Since there are only twelve months, and there is no zeroth month, the appropriate domain is $1 \leq n \leq 12$.

Math.Content.F.IF.6

7. B: The average rate of change of a function $f(x)$ from x_1 to x_2 is given by $\frac{f(x_2)-f(x_1)}{x_2-x_1}$. This formula is the same as the one for slope, rewritten in the context of functions. If $x_1 = 1$ and $x_2 = 5$, the average rate of the given function is given by expression below:

$$\frac{f(x_2) - f(x_1)}{x_2 - x_1} = \frac{f(5) - f(1)}{5 - 1}$$

Calculate the values of $f(1)$ and $f(5)$:

$$f(1) = -3(1) + 1 = -2$$

$$f(5) = -3(5) + 1 = -14$$

Then substitute these values into the expression for the average rate of change and simplify the result:

$$\begin{aligned}\frac{f(5) - f(1)}{5 - 1} &= \frac{(-14) - (-2)}{5 - 1} \\ &= \frac{-12}{4} \\ &= -3\end{aligned}$$

Math.Content.F.IF.7a

8. C: The graph of a linear function is a straight line. To graph the given function, first create a table of values.

x	f(x) = 2x – 4	(x,y)
0	2(0) – 4 = –4	(0,–4)
1	2(1) – 4 = –2	(1,–2)
2	2(2) – 4 = 0	(2,0)

Next plot the three points on a coordinate axis and connect them with a straight line. The *x*- and *y*-intercepts are the points where the line passes through the *x*- and *y*-axes.

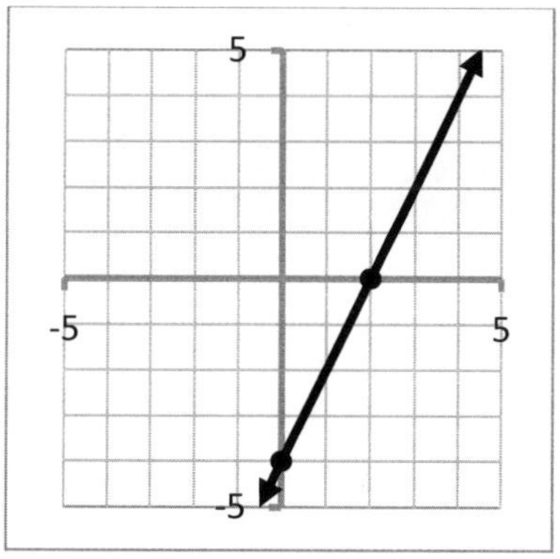

Math.Content.F.IF.7a

9. D: The graph of a quadratic function is a parabolic curve. To graph the given function, first create a table of values.

x	$g(x) = x^2 - 2x - 3$	(x,y)
−2	$(-2)^2 - 2(-2) - 3 = 5$	(−2,5)
−1	$(-1)^2 - 2(-1) - 3 = 0$	(−1,0)
0	$(0)^2 - 2(0) - 3 = -3$	(0,−3)
1	$(1)^2 - 2(1) - 3 = -4$	(1,−4)
2	$(2)^2 - 2(2) - 3 = -3$	(2,−3)

Next plot the five points on a coordinate axis and connect them with a curve. The graph will be a U-shaped parabola. Plot the minimum, which is the lowest point on the graph.

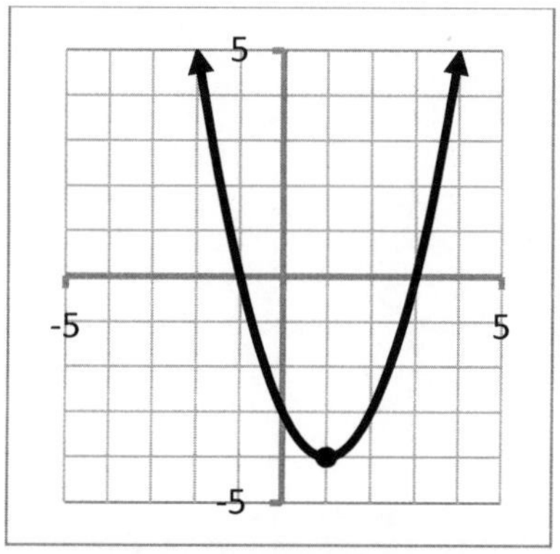

Math.Content.F.IF.7b

10. D: A piecewise-defined function is a function whose definition changes based on certain conditions. In this case, the given function *h* is defined as $\frac{1}{2}x$ when *x* is strictly less than 2 and defined as $4 - x$ otherwise. As a result, the graph is composed of two different straight lines. To graph $h(x)$, first create a table of values for the first half of the function, that is, when x is less than 2:

x	$h(x)$	(x,y)
0	$\frac{1}{2}(0) = 0$	(0,0)
1	$\frac{1}{2}(1) = \frac{1}{2}$	$(1, \frac{1}{2})$
2	$\frac{1}{2}(2) = 1$	(2,1)

Next plot these three points on a coordinate axis. For the point (2,1), plot an open circle since it is not actually on the graph of h (the first half of the function only applies when x is strictly less than 2). Then connect the points to make a line:

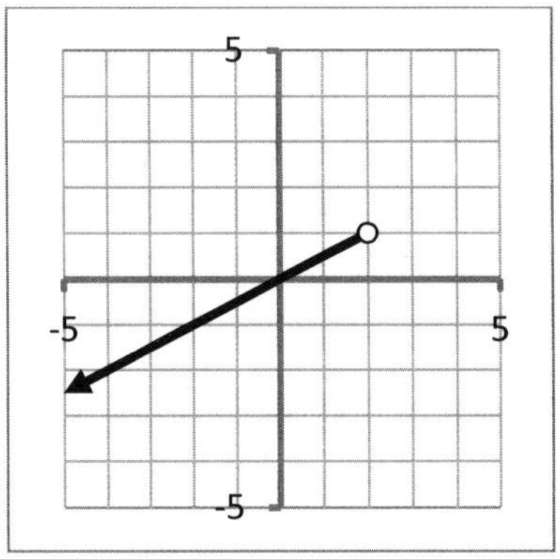

Use a similar procedure to graph the second half of the function. This time use a closed (i.e. filled-in) circle to plot the point at $x = 2$ since it is part of the second half of the function. Graph this part of h on the same graph.

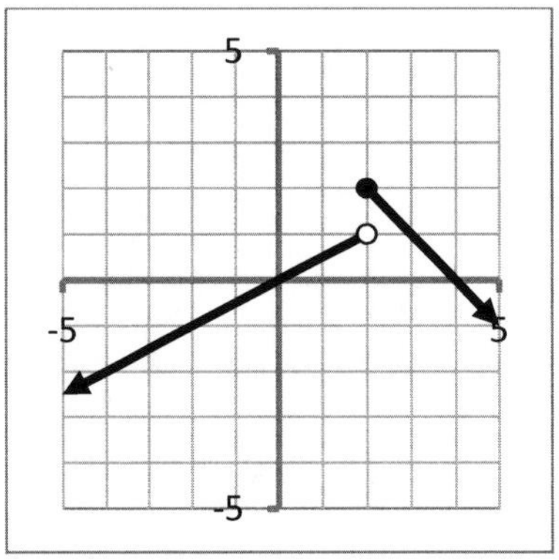

Math.Content.F.IF.7c

11. C: Graph the given function by either creating a table of values by hand or by inputting the function $y = x^3 + x^2 - 4x - 4$ into a graphing calculator.

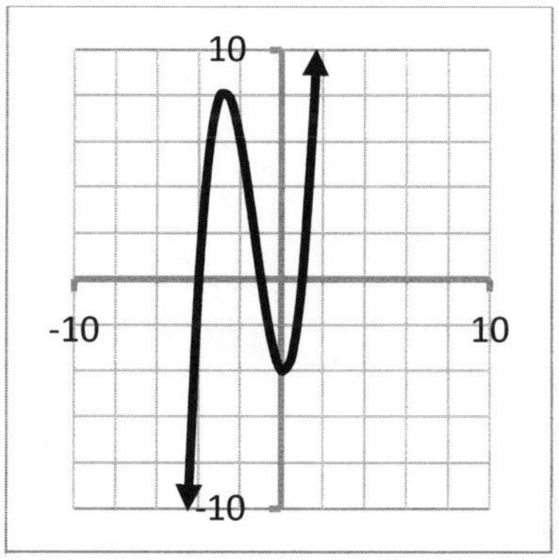

The zeroes of a function are the values where the function equals zero, so in this case, where $f(x) = 0$. You might be able to read these off your table of values. Graphically, the zeroes occur where the graph crosses the x-axis. Notice from the graph above that f crosses the x-axis three times: at the points $(-4,0)$, $(-1,0)$, and $(1,0)$. Plot these points on the graph of f.

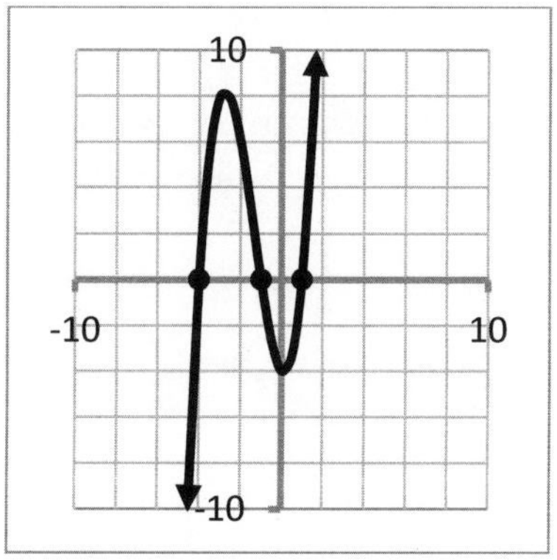

Math.Content.F.IF.7d

12. D: Graph the given function by either creating a table of values by hand or by inputting the function $y = \frac{3}{2x-4}$ into a graphing calculator.

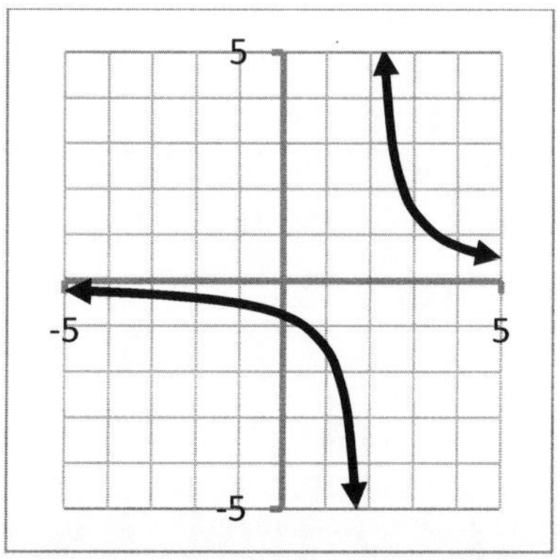

The zeroes of a function are the values where the function equals zero. You might be able to tell from your table of values there are no such points. Graphically, since the graph does not pass through the x-axis at all, g has no zeroes. An asymptote is a line that the graph of a function gets closer and closer to but never touches. Notice that the function never touches the vertical line $x = 2$. (Check: if $x = 2$, $g(2) = \frac{3}{2(2)-4} = \frac{3}{0}$, which is undefined! There can be no value for the function at this value of x.) Therefore, $x = 2$ is a vertical asymptote of g. In addition, g approaches but never touches the horizontal line $y = 0$. (Check: there is no solution for $g(x) = 0$.) Thus, the second asymptote is $y = 0$. Graph these lines on the same graph with light, dotted lines.

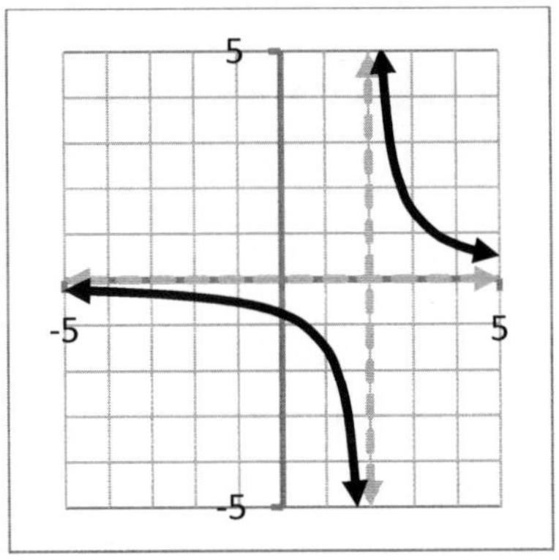

Math.Content.F.IF.7e

13. A: Graph the given function by either creating a table of values by hand or by inputting the function $y = \log_2 2(x + 3)$ into a graphing calculator. In addition, plot the points where the line passes through the x- and y-axes, which are the x- and y-intercepts of h.

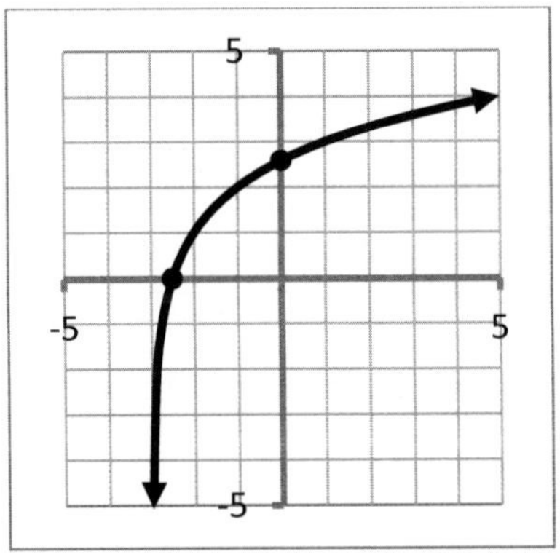

Math.Content.F.IF.7e

14. C: Graph the given function by either creating a table of values by hand or by inputting the function $y = 2\sin(3x - \frac{\pi}{2})$ into a graphing calculator. The graph of a trigonometric function like sine is a periodic function. That is, it repeats its values after a certain interval.

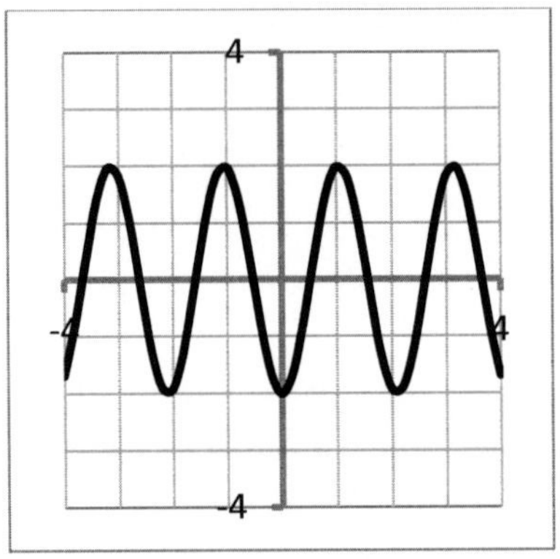

Math.Content.F.IF.8a

15. B: The zeroes of a function are the domain values where the function equals zero. So in this case, the zeroes are the values of x for which $f(x) = 0$. To find the zeroes of the function, set up an equation and then solve it for x. The first step is to factor the quadratic expression:

$$f(x) = 0$$
$$x^2 + 5x - 24 = 0$$
$$(x + 8)(x - 3) = 0$$

When the coefficient of x^2 is 1, factor the quadratic into a product of binomials $(x + a)(x + b)$, where a and b are chosen so their product is the constant term in the quadratic (including sign), and their sum is the coefficient of the linear term in the quadratic (the x term). Then use the zero-product rule to solve the result. Using that factorization of f, write two equations and solve them both for x:

$$x + 8 = 0 \text{ or } x - 3 = 0$$

$$x = -8 \text{ or } x = 3$$

Therefore, the zeroes of the function are –8 and 3.

Math.Content.F.IF.8b

16. B: In general, if an initial investment of P dollars earns interest at r percent (expressed as a decimal) per year, then its value V after t years is given by the function $V = P(1 + r)^t$. The given function is $V = 200(1.04)^t$, so the percent interest is 0.04, or 4%, per year.

Math.Content.F.IF.9

17. B: The x-intercept of a function is the point where the line passes through the x-axis. Since the graph passes through the x-axis at (6,0), its x-intercept is 6. To determine which function has the same x-intercept, substitute 6 for x into each function and see if the result is zero. Start with choice A, $g(x) = -4x - 12$.

$$\begin{aligned} -4(6) - 12 &= -24 - 12 \\ &= -36 \end{aligned}$$

Since the result is not 6, the x-intercept is not 6. Next try choice B, $g(x) = x^2 - 12x + 36$.

$$\begin{aligned} (6)^2 - 12(6) + 36 &= 36 - 72 + 36 \\ &= 0 \end{aligned}$$

Thus, the function $g(x) = x^2 - 12x + 36$ has the same x-intercept as the graphed function.

Math.Content.F.BF.1a

18. B: Since the theater will sell 500 tickets if it charges $10 per ticket, we know that $t(10) = 500$. Furthermore, because of the way ticket price affects ticket sales, t must be a linear function that decreases by 50 every time d increases by 1. Therefore, the d-term of the function is $-50d$, so the function takes the form $t(d) = -50d + c$. To find the value of c, substitute 10 for d and 500 for $t(10)$ and solve for c.

$$t(10) = -50(10) + c$$

$$\begin{aligned} 500 &= -500 + c \\ 1000 &= c \end{aligned}$$

Thus, the function is $t(d) = -50d + 1000$.

Math.Content.F.BF.1b

19. D: The cost of the taxi ride is the sum of two functions, a constant function for the first mile and a linear function for the rest of the ride. The constant function is $c_1(d) = 4.25$ since the cost of the first mile is $4.25. For the linear part, subtract 1 from d to exclude the first mile, and then multiply the result by 0.70 since it costs $0.70 per mile. The result is $c_2(d) = 0.70(d - 1)$. Finally, write the function for the total cost of the taxi ride by adding the two functions.

$$\begin{aligned} c(d) &= c_1(d) + c_2(d) \\ &= 4.25 + 0.70(d - 1) \end{aligned}$$

Math.Content.F.BF.1c

20. B: The first function $L(g)$ gives the water level after g gallons are pumped into the tank. The second function $w(t)$ gives the number of gallons pumped into the tank after t minutes, which the first function calls g. Consequently, we can have L act on w: the composition of the functions $L(w(t))$ is the water level of the tank after t minutes. Calculate $L(w(t))$.

$$\begin{aligned} L(w(t)) &= 0.3 \cdot w(t) \\ &= 0.3 \cdot 1.2t \\ &= 0.36t \end{aligned}$$

Thus, the function $L(t) = 0.36t$ represents the water level of the tank after t minutes.

Math.Content.F.BF.2

21. D: The number of wheat grains on each square forms a geometric sequence: 1, 2, 4, 8, 16, etc. Notice that these numbers are all powers of two (i.e. $2^0, 2^1, 2^2, 2^3, 2^4$, etc.). Therefore, the sequence can be written as an exponential function in the form $y = 2^x$. Moreover, since this sequence begins (at square number 1) with an exponent of 0 rather than 1, we need to subtract 1 from n to get the correct exponent of 2. Thus, the function $g(n) = 2^{n-1}$ represents the number of wheat grains on the nth square of the chessboard.

Math.Content.F.BF.3

22. C: To transform $f(x)$ into $g(x)$, you have to replace x with $x + 1$. This transformation results in a translation, or shift, of the graph. Specifically, for a particular x, $g(x)$ returns the value of the function $f(x)$ one unit further along the x-axis (to the right). This results in shifting the graph one unit to the left.

Math.Content.F.BF.3

23. D: To transform $f(x)$ into $g(x)$, you have to multiply it by 5. This transformation increases the distance between y-values of the same two x values, and so we see it stretches the graph vertically, making it five times "taller".

Math.Content.F.BF.4a

24. D: The inverse of a function is another function that *undoes* the original one. By convention, the inverse of $f(x)$ is written as $f^{-1}(x)$. To find the inverse of $f(x)$, first replace $f(x)$ with y. Then switch all the x's with all the y's (and vice versa):

$$f(x) = \frac{3-x}{2} \quad \text{(original)}$$
$$y = \frac{3-x}{2} \quad \text{(replaced f(x) with y)}$$
$$x = \frac{3-y}{2} \quad \text{(exchanged x and y)}$$

Next solve the equation for y:

$$2x = 3 - y$$
$$y + 2x = 3$$
$$y = 3 - 2x$$

Finally, replace y with $f^{-1}(x)$ to write the result as an inverse function of f(x):

$$f^{-1}(x) = 3 - 2x$$

Math.Content.F.BF.4a

25. C: The inverse of a function is another function that *undoes* the original one. By convention, the inverse of $f(x)$ is written as $f^{-1}(x)$. To find the inverse of $f(x)$, first replace $f(x)$ with y. Then switch all the x's with all the y's (and vice versa):

$$f(x) = x^3 + 8 \quad \text{(original)}$$
$$y = x^3 + 8 \quad \text{(replaced f(x) with y)}$$
$$x = y^3 + 8 \quad \text{(exchanged x and y)}$$

Next solve the equation for y:

$$x - 8 = y^3$$
$$\sqrt[3]{x-8} = \sqrt[3]{y^3}$$
$$\sqrt[3]{x-8} = y$$

Finally, replace y with $f^{-1}(x)$ to write the result as an inverse function of f(x):

$$f^{-1}(x) = \sqrt[3]{x-8}$$

Math.Content.F.BF.4b

26. C: The inverse of a function is another function that *undoes* the original one. For a function f^{-1} to be the inverse of f, their composition must be equal to x as shown below.

$$f^{-1}(f(x)) = x$$
$$f(f^{-1}(x)) = x$$

Therefore, the composition of the given functions, $\cos^{-1} t$ and cos t, is the variable t.

$$\cos^{-1}(\cos t) = t$$

Math.Content.F.BF.4c

27. B: The inverse of a function is another function that *undoes* the original one. By convention, the inverse of $g(x)$ is written as $g^{-1}(x)$. For any value of x, the result of inputting x into g can itself be inputted into g^{-1}, and it will return the original value x. In other words, if $g(x) = y$, then $g^{-1}(y) = x$.

In regard to the problem, this means that the value of $g^{-1}(3)$ is the value of x for which $g(x) = 3$. Examining the graph of g, notice that it passes through the point (1,3). Therefore, $g(1) = 3$, and the value of $g^{-1}(3)$ is 1.

Math.Content.F.BF.4c

28. B: The inverse of a function is another function that *undoes* the original one. By convention, the inverse of $h(x)$ is written as $h^{-1}(x)$. For any value of x, the result of inputting x into h can itself be inputted into h^{-1}, and it will return the original value x. In other words, if $h(x) = y$, then $h^{-1}(y) = x$.

In regard to the problem, this means that the value of $h^{-1}(-2)$ is the value of x for which $h(x) = -2$. Examining the table for h, notice that when $x = -1$, $h(x) = -2$. Therefore, $h(-1) = -2$, and the value of $h^{-1}(-2)$ is -1.

Math.Content.F.BF.4d

29. B: A function is invertible if its inverse is a function. The given function f is not invertible if the domain is all real numbers because in some cases different values of x result in the same value of $f(x)$. Graphically, this means that the function does not pass the horizontal line test. That is, at least one horizontal line (like $y = 3$) intersect the graph at more than one point (the points (0,3) and (4,3)).

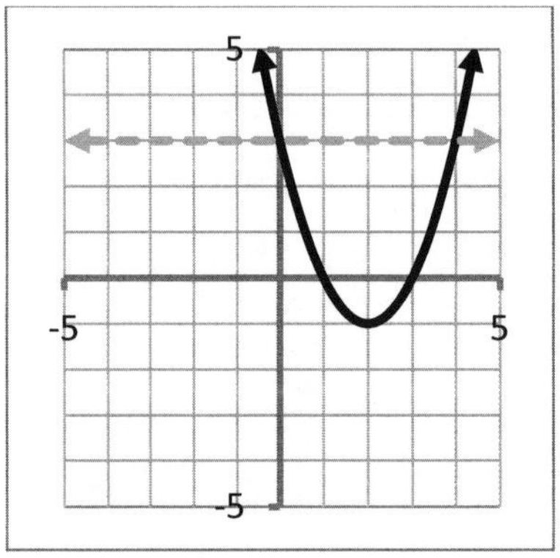

To make the function invertible, restrict the domain so that the new function does pass the horizontal line test. If you restrict the domain to $x \geq 2$, the resulting function will pass the horizontal line test as shown in the graph below.

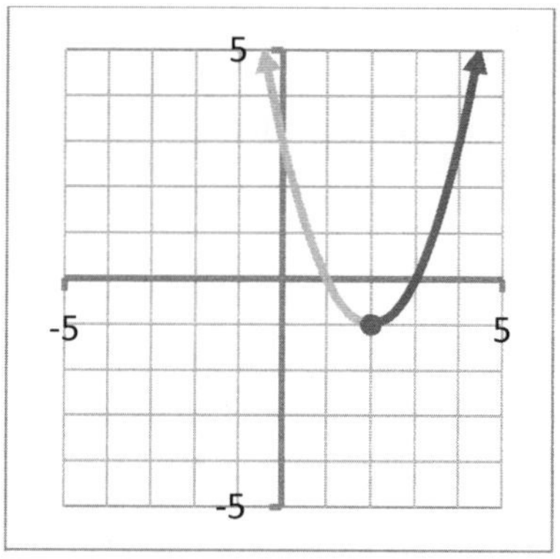

Math.Content.F.BF.5

30. B: The logarithm of a number is the exponent that the base must to be raised to in order to get that number. For example, since $2^4 = 16$, it is also true that $\log_2 16 = 4$. By this definition, the given equation is equivalent to the following equation:

$$x^2 - 8 = 2^3$$

Simplify this exponential equation and solve for x:

$$\begin{aligned} x^2 - 8 &= 8 \\ x^2 &= 16 \\ x &= \pm 4 \end{aligned}$$

Thus, the solutions are $x = -4$ and $x = 4$.

Math.Content.F.LE.1a

31. B: The length of an interval is the difference between its endpoints. For example, the length of the interval [2, 4] is 2. To determine how the given function grows over an interval of length 3, determine the value of f at each endpoint of that interval. Since exponential functions grow by equal factors over equal intervals, you can use any interval of length 3, and your answer will apply to all such intervals. For example, you can use the interval [0,3]:

$$\begin{aligned} f(0) &= 3 \cdot 2\text{^}((0)) \\ &= 3 \cdot 1 \\ &= 3 \\ f(3) &= 3 \cdot 2\text{^}((3)) \\ &= 3 \cdot 8 \\ &= 24 \end{aligned}$$

Since $f(0) = 3$ and $f(3) = 24$, the function grows by a factor of $\frac{24}{3} = 8$ over this interval.

Math.Content.F.LE.1b

32. C: Linear functions grow by equal differences (rather than equal factors) over equal intervals. In other words, if the linear function $c(f)$ converts a temperature f from Fahrenheit to Celsius, then intervals of equal length (in f) result in equal increases in the value of the function $c(f)$.

From the problem, we know that $c(32) = 0$ and $c(68) = 20$. Thus, we can conclude that intervals of length 36 (like the interval [32,68]) result in an increase of 20 since $20 - 0 = 20$. In addition, since the length of [68,104] is 36, the function $c(f)$ increases by 20 over this interval as well. Use this information to calculate $c(104)$.

$$c(104) = c(68) + 20 = 20 + 20 = 40$$

Therefore, we can conclude that 104°F is equivalent to 40°C.

Math.Content.F.LE.1c

33. D: Exponential functions grow by equal factors over equal intervals. In other words, if the exponential function $n(t)$ represents the number of bacteria in the Petri dish after t days, then intervals of equal length (in t) multiply the value of the function $n(t)$ by equal factors.

From the problem, we know that $n(0) = 160$, $n(2) = 360$, and $n(3) = 540$. Use division to determine how intervals of length 2 (like the interval [0,2]) change the value of $n(t)$.

$$\frac{n(2)}{n(0)} = \frac{360}{160}$$

$$2.25$$

Therefore, increasing t by 2 multiplies the value of $n(t)$ by 2.25. Similarly, since the length of [3,5] is 2, the interval [3,5] multiplies the value of $n(t)$ by 2.25 as well. Use this information to calculate $n(5)$:

$$n(5) = n(3) \cdot 2.25 = 540 \cdot 2.25 = 1215$$

Therefore, we can conclude that there are 1215 bacteria in the Petri dish in five days.

Math.Content.F.LE.2

34. D: Linear functions can be written in the form $g(x) = mx + b$. To determine the value of the slope, m, notice that $g(x)$ decreases by 3 every time x increases by 1: when x goes from 0 to 1, the value of the function goes from 4 to 1; and when x goes from 1 to 2, the function goes from 1 to –2. Since $g(x)$ decreases by 3, the slope is –3, so the function is $g(x) = -3x + b$.

Next, calculate the value of the y-intercept, b. The y-intercept is the value of y, or in this case $g(x)$, when $x = 0$. From the table, you can tell that $g(0) = 4$. Thus, the function is $g(x) = -3x + 4$.

Math.Content.F.LE.2

35. B: Exponential functions can be written in the form $h(x) = ab^x$. To determine the value of b, notice that $h(x)$ doubles every time x increases by 1. For instance, when x goes from –2 to –1, the value of y goes from 1 to 2; and when x goes from –1 to 0, the value of y doubles from 2 to 4. Since the graph doubles, the value of b is 2, so the function is $h(x) = a \cdot 2^x$. Next calculate the value of a. From the graph, you can tell that $h(0) = 4$. Use this to calculate a:

$$\begin{aligned} h(0) &= 4 \\ a \cdot 2^0 &= 4 \\ a \cdot 1 &= 4 \\ a &= 4 \end{aligned}$$

Thus, the function is $h(x) = 4 \cdot 2^x$.

Alternatively, from the general form of an exponential $h(x) = ab^x$, notice that if $x = 0$, $h(x) = a$. Therefore, since $y = 4$ when $x = 0$, $a = 4$. You can solve for b from any of the other indicated points, for example:

$$\begin{aligned} 2 &= 4b - 1 = 4/b \\ b &= 2 \end{aligned}$$

Therefore, the function is $h(x) = 4 \cdot 2^x$.

Math.Content.F.LE.3

36. A: Examine the graph to determine which graph eventually exceeds the other three. In other words, find the function that eventually goes above the other three as the graph goes from left to right. Notice that the function f goes above the other three when $x = 5$.

Math.Content.F.LE.4

37. A: You will need to use the natural logarithm, ln, to solve this equation. First divide both sides of the equation by 3:

$$3e^{2t} = 12$$

$$e^{2t} = 4$$

Next take the natural logarithm of both sides to get rid of the exponent. The natural logarithm is defined as the logarithm to the base e; the natural log is the inverse function of the exponential. Therefore, $\ln e^x = x$.

$$\ln e^{2t} = \ln 4$$

$$2t = \ln 4$$

Finally, divide both sides by 2 to solve for t.

$$t = \frac{\ln 4}{2}$$

Math.Content.F.LE.5

38. C: The linear function $s(h) = 27h + 720$ starts at 720 when $h = 0$ and increases by 27 every time h increases by 1. Therefore, for every hour after 40, the librarian earns \$27, so his salary is \$27 per hour.

Math.Content.F.LE.5

39. D: The general form of the interest exponential is $A(t) = P(1 + r)^t$, where P is the initial principal, r the interest rate (as a decimal), and t the number of interest yield periods after the initial investment. P is \$315 in this case, which means that is the initial amount put into the account. Furthermore, the exponential function $b(t) = 315(1.05)^t$ equals 315 when $t = 0$. Therefore, John initially put \$315 into the bank account.

Math.Content.F.TF.1

40. B: An arc is a piece of a circle. In the figure, $\widehat{AB}$ is the piece of the circle that starts at point A and ends at B. In general an arc length s is given by $s = \theta R$, where R is the radius of the circle containing the arc and θ is the angle subtended by radii drawn to the endpoints of the arc. In a unit circle, the length of an arc is simply the measure of the angle (in radians) subtended by the angle. Therefore, the arc length of $\widehat{AB}$ is equal to the measure of $\angle AOB$, so its length is $\frac{\pi}{3}$.

Math.Content.F.TF.2

41. A: In a unit circle, trigonometric functions can be represented by considering angles that start from the positive side of the *x*-axis and are measured counterclockwise around the circle. For angles like this, the sine of the angle is the *y*-coordinate of the point where the angle's side meets the unit circle. Since one side of the angle meets the unit circle at (–0.6, –0.8), the value of sin θ is –0.8.

Math.Content.F.TF.3

42. B: The angle π/6 is in radians. To convert it to degrees, multiply by $\frac{180}{\pi}$.

$$\frac{\pi}{6} \cdot \frac{180}{\pi} = 30°$$

Therefore, $\tan\frac{\pi}{6} = \tan 30°$. To calculate this value, draw a 30-60-90 triangle, which is a special triangle whose proportions you may have memorized. Making the hypotenuse one unit long simplifies things, although it is not necessary as long as the proportions are the same.

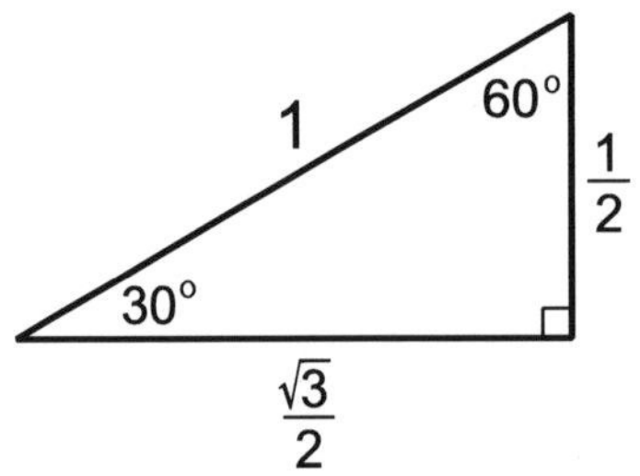

By SOH-CAH-TOA, the tangent function is defined as $\frac{\text{opposite}}{\text{adjacent}}$ in a right triangle. Therefore, the value of tan 30° is $\frac{1/2}{\sqrt{3}/2}$. Simplify this fraction:

$$\begin{aligned}\tan\frac{\pi}{6} &= \frac{1/2}{\sqrt{3}/2} \\ &= 1/\sqrt{3} \\ &= \sqrt{3}/3\end{aligned}$$

Alternatively, remember that tan θ = (sin θ) / (cos θ). You might have memorized the sine and cosine of 30° as ½ and √3/2, respectively. This is the same division as above, and yields the same answer.

Math.Content.F.TF.3

43. D: All trigonometric functions can be visualized with a unit circle on a coordinate plane. If an angle is measured from the positive side of the *x*-axis in the counterclockwise direction, then the sine function is the *y*-coordinate of the point where the angle's other side meets the unit circle. To visualize the problem, draw a unit circle on a coordinate plane showing an arbitrary angle *x* and the angle $\pi - x$.

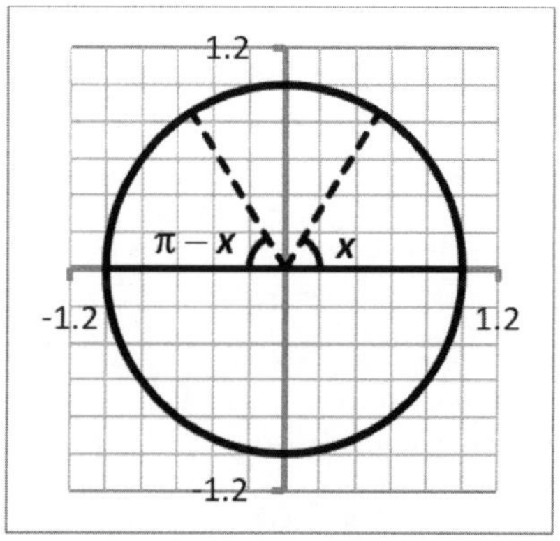

Notice that the *y*-coordinate associated with the angles *x* and $\pi - x$ are equal. Therefore, sine of *x* and sine of $\pi - x$ are equal, so $\sin(\pi - x) = \sin x$.

Math.Content.F.TF.4

44. C: Graph the trigonometric function $y = \cos x$.

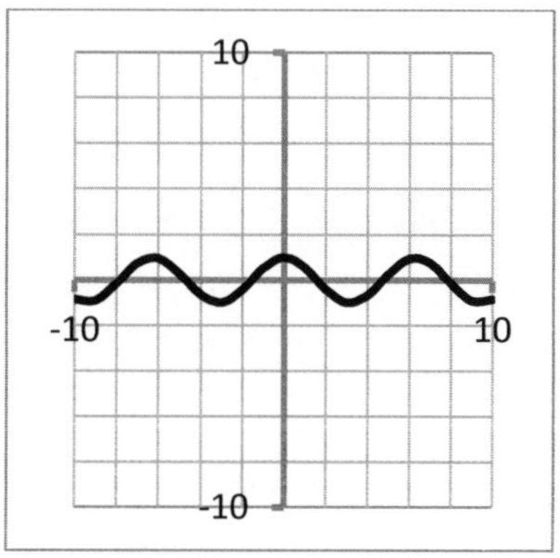

A function is periodic if it repeats its value in regular intervals and aperiodic if it doesn't. Since the graph of $y = \cos x$ repeats itself it is periodic.

A function has even symmetry if it is symmetric about the *y*-axis and odd symmetry if it can be rotated 180° about the origin and look the same. Since the graph has symmetry about the *y*-axis, it has even symmetry. Furthermore, you might see that for a given angle, say 45°, replacing it with -45° does not change the x-component of the point of intersection with the unit circle, so that cos (-*x*) = cos (x), which is even symmetry.

Math.Content.F.TF.5

45. D: In 1 s, the weight returns to its previous height, and so the period T of the function is 1. The frequency of a periodic function is the number of cycles of the function per unit time, and is the reciprocal of the period. Thus, the frequency of h is 1. The amplitude is half the distance between the lowest and highest point: $A = ½[5, 25] = 10$. The midline is given by the lowest point plus the amplitude or the highest minus the amplitude: $5 + 10 = 25 - 10 = 15$.

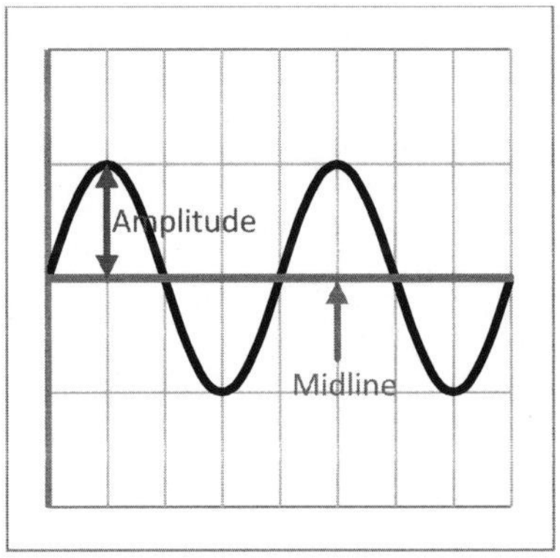

Now write the function $h(t)$ by modifying the parent function $y = \cos t$. The amplitude of $\cos t$ is 1. To make the amplitude 10, multiply the parent function by 10.

$$y = 10cos\ t$$

The midline of this new function is 0. To move the midline to 15, add 15.

$$y = 10cos\ t + 15$$

This function repeats once every 2π, so its period is 2π, and its frequency is therefore $\frac{1}{2\pi}$. To make the frequency 1, multiply t by 2π so the function repeats every second.

$$h(t) = 10cos(2\pi t) + 15$$

At $t = 0$, the weight is at its lowest point, instead of its highest (where the parent function $\cos t$ starts). To start $h(t)$ at its lowest point, there are two options. Either multiply the amplitude by -1:

$$h(t) = -10cos(2\pi t) + 15$$

Or offset the argument by 180°:

$$h(t) = 10cos(2\pi t - \pi) + 15$$

Since the latter is not shown, choice D is correct.

Math.Content.F.TF.6

46. A: A function is invertible if its inverse is a function. The given function f is not invertible if the domain is all real numbers because, in some cases, different values of x result in the same value of

$f(x)$. Graphically, this means that the function does not pass the horizontal line test. That is, at least one horizontal line (like $y = 0.5$) intersect the graph at more than one point.

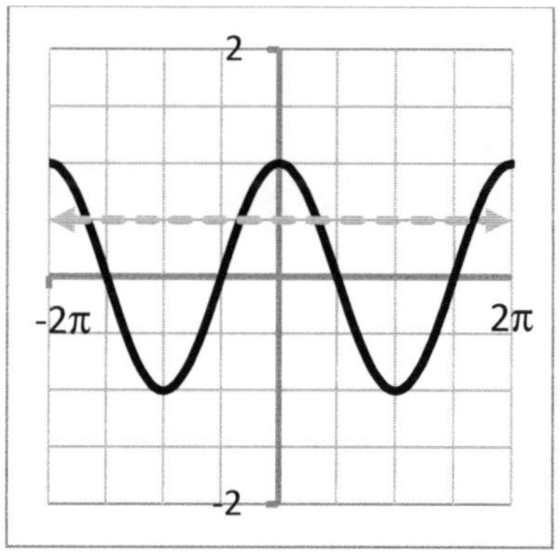

To make the function invertible, restrict the domain so that the new function does pass the horizontal line test. The only choice that accomplishes this is A: if you restrict the domain to $0 \leq x \leq \pi$, the resulting function will pass the horizontal line test as shown in the graph below.

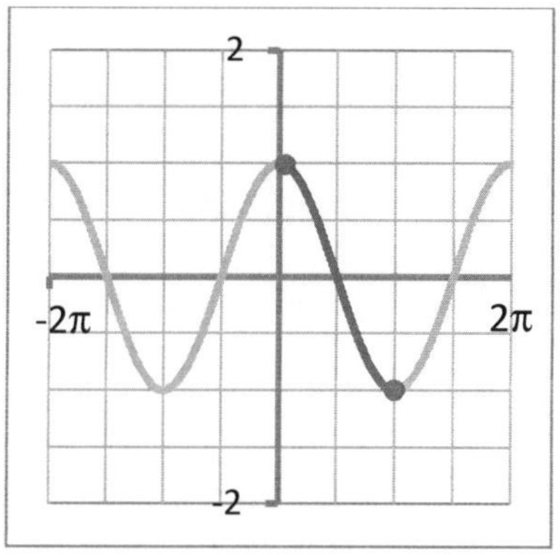

Math.Content.F.TF.7

47. B: First set up an equation for when John's height is 90 m.

$$h(t) = 90$$
$$-60\,cos\left(\frac{\pi t}{15}\right) + 70 = 90$$

To solve the equation, isolate $\cos\left(\frac{\pi t}{15}\right)$:

$$-60\cos\left(\frac{\pi t}{15}\right) = 20$$
$$\cos\left(\frac{\pi t}{15}\right) = -\frac{1}{3}$$

Then take the inverse cosine of both sides of the equation:

$$\cos^{-1}\left[\cos\left(\frac{\pi t}{15}\right)\right] = \cos^{-1}\left(-\frac{1}{3}\right)$$
$$\frac{\pi t}{15} = \cos^{-1}\left(-\frac{1}{3}\right)$$

Use a calculator to estimate the right side (in radians):

$$\frac{\pi t}{15} \approx 1.910633$$

Finally, solve for *t*.

$$t \approx \frac{15(1.910633)}{\pi}$$
$$t \approx 9$$

Therefore, John is 90 m above the ground after about 9 minutes.

Math.Content.F.TF.8

48. C: The Pythagorean identity is shown below.

$$\sin^2\theta + \cos^2\theta = 1$$

To make it easier to use in this problem, an equivalent identity can be written by subtracting $\sin^2\theta$ from both sides.

$$\cos^2\theta = 1 - \sin^2\theta$$

Substitute $\frac{4}{5}$ for $\sin\theta$ into this equation and solve for $\cos^2\theta$.

$$\cos^2\theta = 1 - \sin^2\theta$$
$$= 1 - \left(\frac{4}{5}\right)^2$$
$$= 1 - \frac{16}{25}$$
$$= \frac{9}{25}$$

Next take the square root of each side. Since θ is in quadrant I, $\cos\theta$ is positive.

$$\cos^2\theta = \frac{9}{25}$$

$$\cos\theta = \frac{3}{5}$$

Since $\tan\theta = \frac{\sin\theta}{\cos\theta}$, you can calculate $\tan\theta$ by dividing $\sin\theta = \frac{4}{5}$ by $\cos\theta = \frac{3}{5}$. Simplify the result.

$$\tan\theta = \frac{\sin\theta}{\cos\theta}$$
$$= \frac{4/5}{3/5}$$
$$= \frac{4}{3}$$

You might also have memorized the Pythagorean triple {3, 4, 5} as the sides of a right triangle. If the sine of an angle in the first quadrant is 4/5, then the side of the right triangle adjacent to it is 3. Since $\tan\theta = \frac{opposite}{adjacent}$, tan θ = 4/3.

Math.Content.F.TF.9

49. D: The addition formula for sine is shown below.

$$\sin(\alpha + \beta) = \sin\alpha\cos\beta + \cos\alpha\sin\beta$$

Since you know the values of sin θ and cos θ when θ = 30° and when θ = 45°, substitute these values for α and β into the addition formula. Then simplify the result:

$$sin\ 75° = sin\ (30° + 45°) = sin\ 30° cos\ 45° + cos\ 30° sin\ 45°$$

$$\sin 75° = \frac{1}{2}\cdot\frac{\sqrt{2}}{2} + \frac{\sqrt{3}}{2}\cdot\frac{\sqrt{2}}{2}$$

$$= \frac{\sqrt{2}}{4} + \frac{\sqrt{6}}{4}$$

$$= \frac{\sqrt{2} + \sqrt{6}}{4}$$

Math.Content.F.TF.9

50. A: The subtraction formula for tangent is shown below.

$$\tan(\alpha - \beta) = \frac{\tan\alpha - \tan\beta}{1 + \tan\alpha\tan\beta}$$

To calculate tan $(x - y)$, substitute the values of tan x and tan y into the formula. Then simplify the result:

$$\tan(x - y) = \frac{\tan x - \tan y}{1 + \tan x \tan y}$$

$$= \frac{{}^{4}/_{3} - {}^{1}/_{2}}{1 + ({}^{4}/_{3})({}^{1}/_{2})}$$

$$= \frac{5/6}{1 + (2/3)}$$

$$= \frac{5/6}{5/3} = \frac{5}{6} \times \frac{3}{5} = \frac{3}{6}$$

$$= \frac{1}{2}$$

Practice Test #2

Practice Questions

1. The function f is defined as $f(x) = x^2 - 2$. Its domain is all real numbers, and its range is $f(x) \geq -2$. Which of the following is true of f?

a. If $x < -2$, then $f(x)$ is undefined.
b. It is defined when x is negative.
c. The range of the function is $f(-2)$.
d. The value of $f(-2)$ is undefined.

2. If $g(x) = 2x^2 - x + 3$, evaluate g(4).

a. $g(4) = 7$
b. $g(4) = 31$
c. $g(4) = 63$
d. $g(4) = 67$

3. The function $V(r) = \frac{4}{3}\pi r^3$ gives the volume of a sphere of radius r. What is the volume of a sphere of radius 3?

a. 12π
b. 27π
c. 36π
d. 108π

4. Calculate the sixth term of the sequence defined as $f(0) = f(1) = 1; f(n+1) = f(n) + 2 \cdot f(n-1)$ for $n \geq 2$.

a. 8
b. 11
c. 21
d. 41

5. The height of a pendulum (in cm) t seconds after it is released is given by the function $h(t)$. The table below displays the value of h for eight different values of t.

t	0	0.2	0.4	0.6	0.8	1	1.2	1.4
h(t)	10	8	4	2	4	8	10	8

How many seconds after it is released does the pendulum reach its lowest recorded point?

a. After 0.4 seconds
b. After 0.6 seconds
c. After 2 seconds
d. After 4 seconds

6. Which of the following would be an appropriate domain for the function *P*(*t*), the world population (in billions) in the year *t* from 1900 to 2000?

a. $1.5 \leq t \leq 7$
b. $1900 \leq t \leq 2000$
c. $1.5 \leq P(t) \leq 7$
d. $1900 \leq P(t) \leq 2000$

7. Calculate the average rate of change of the function $f(x) = 4x - 7$ from $x = 0$ to $x = 6$.

a. -4
b. $-\frac{5}{3}$
c. $\frac{5}{3}$
d. 4

8. Graph the function $f(x) = -3x + 3$. Show its *x*- and *y*-intercepts.

a.
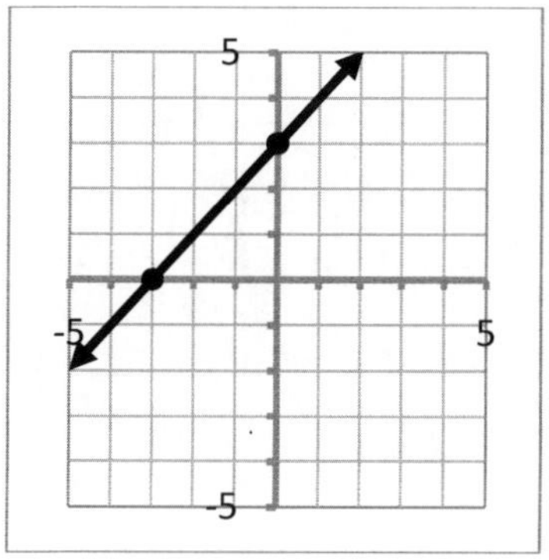

b.
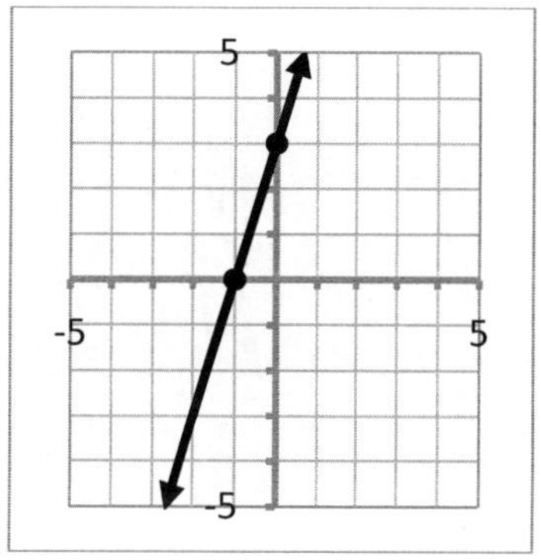

c.
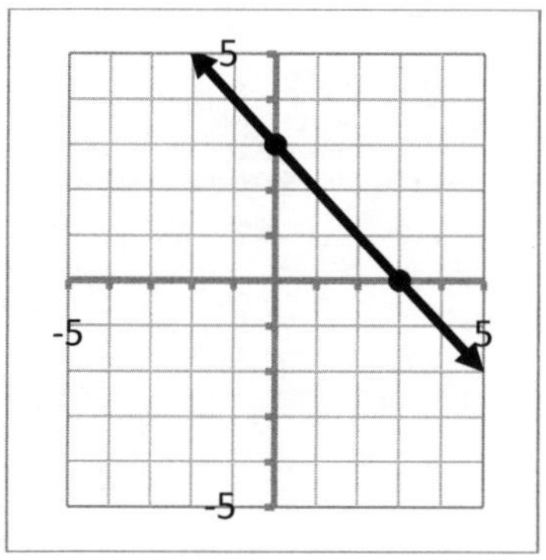

d.
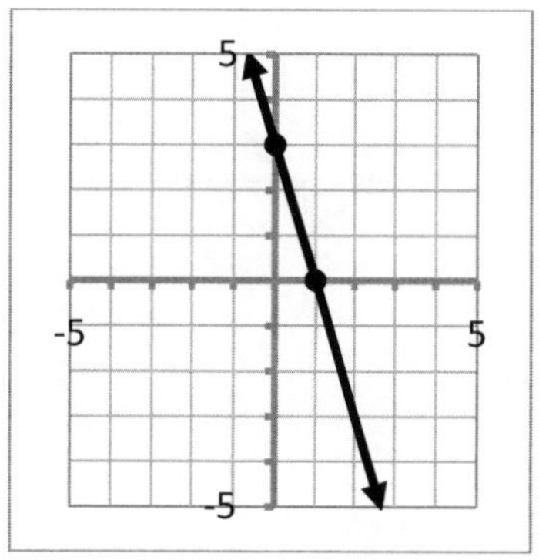

9. Graph the function $g(x) = -x^2 + 4x - 3$. Show its maximum.

a.

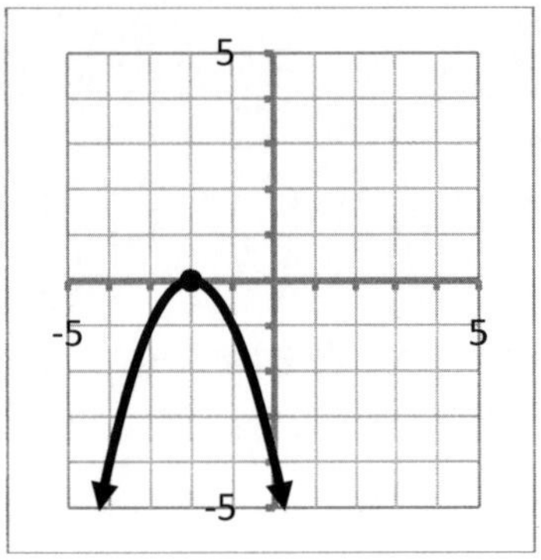

b.

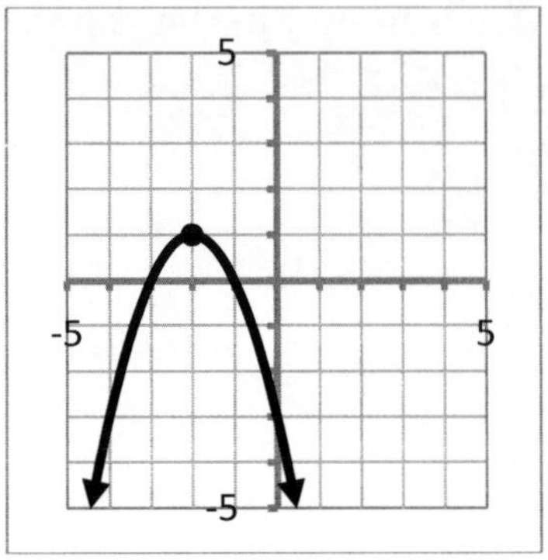

c.

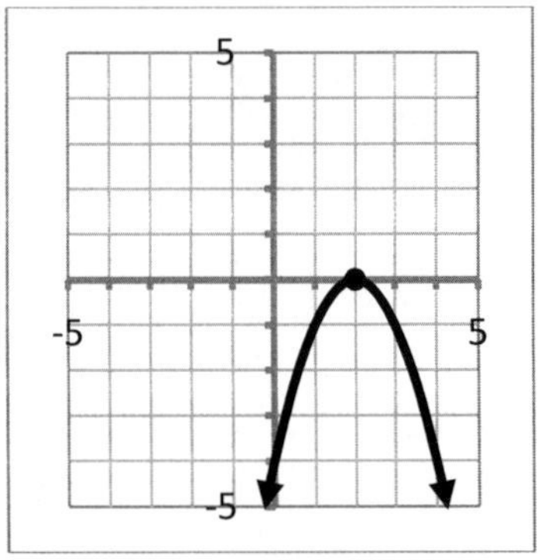

d.

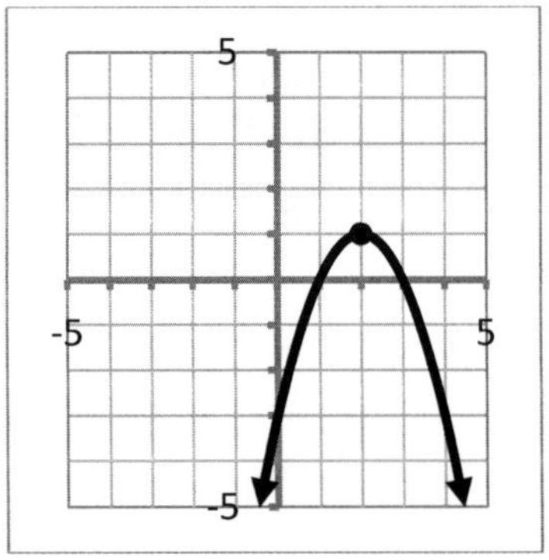

10. Graph the function $h(x) = |x + 2| - 1$.

a.

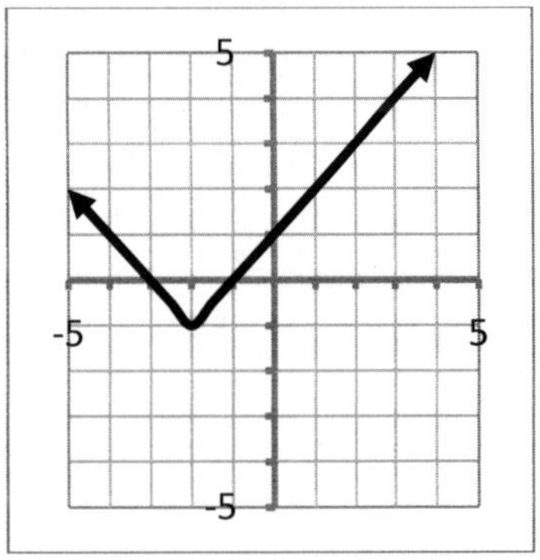

b.

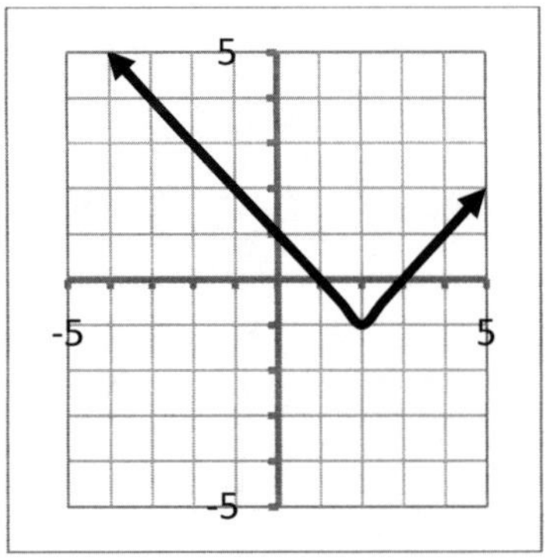

c.

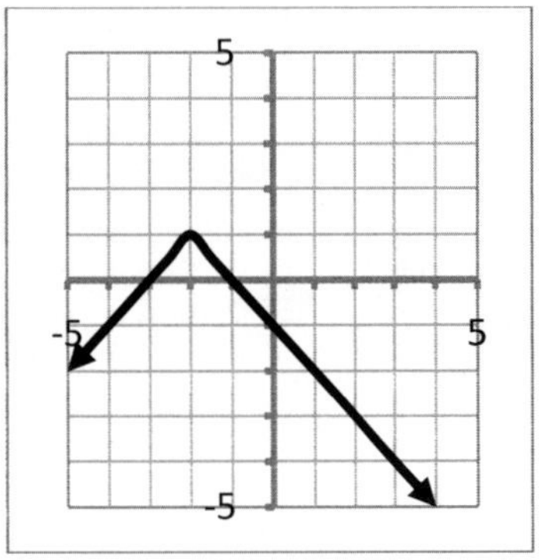

d.

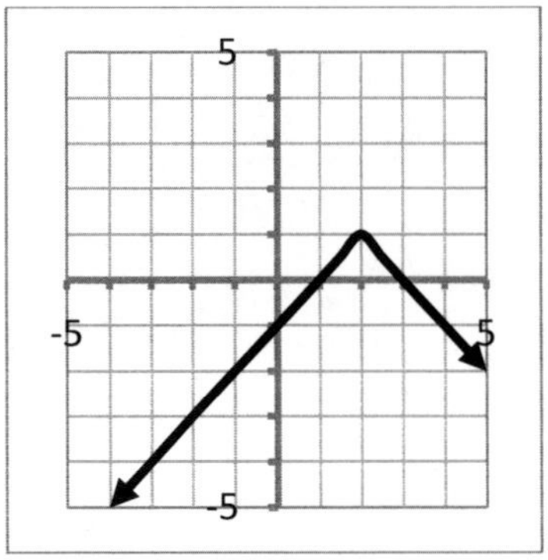

11. Graph the function $f(x) = x^3 - 2x^2 - 4x + 8$. Show the function's zeroes.

a.

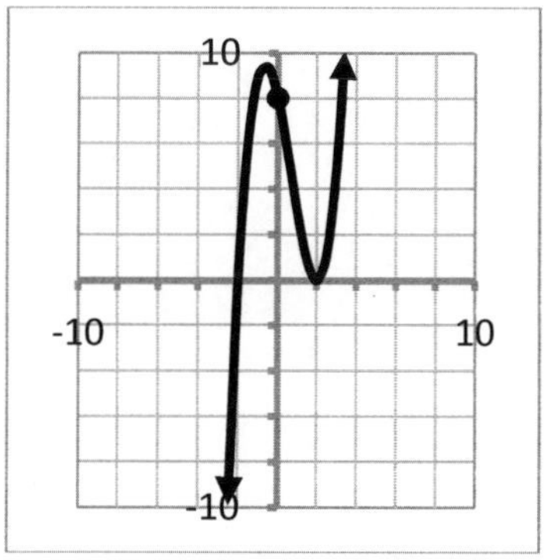

b.

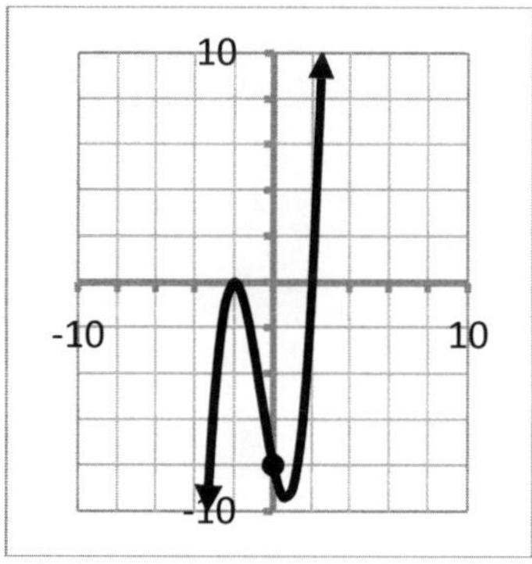

c.

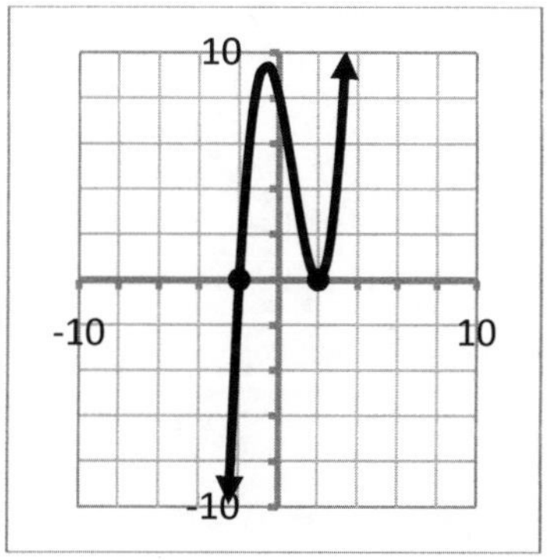

d.

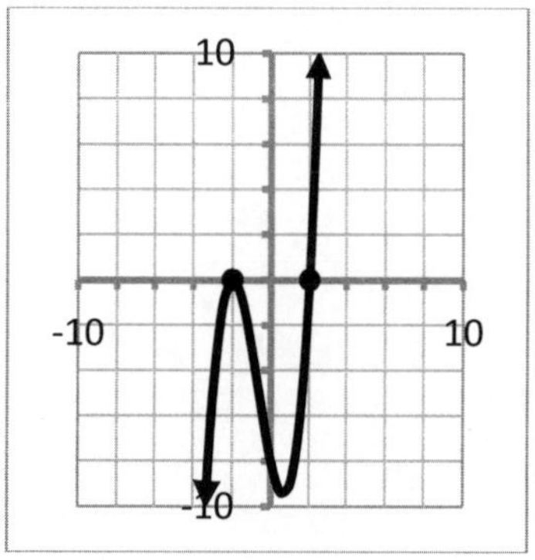

12. Graph the function $g(x) = \frac{3x}{2x^2-8}$. Show its zeroes and asymptotes if there are any.

a.

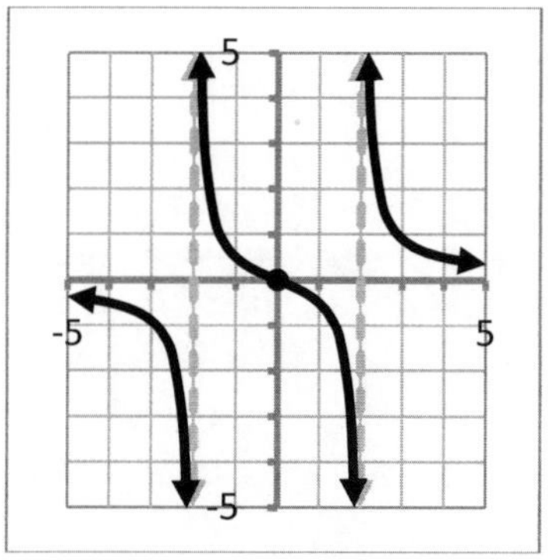

b.

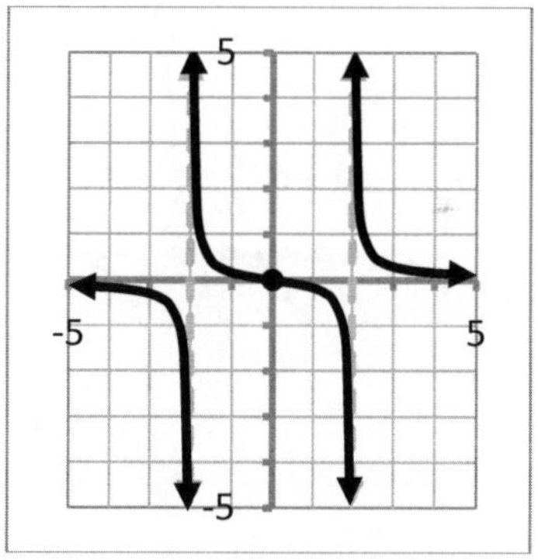

c.

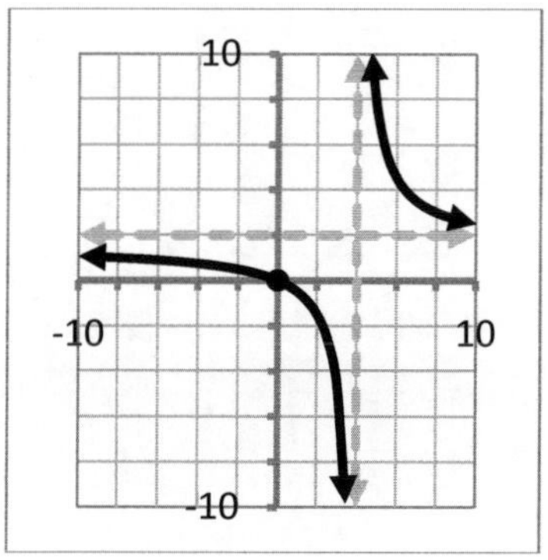

d.

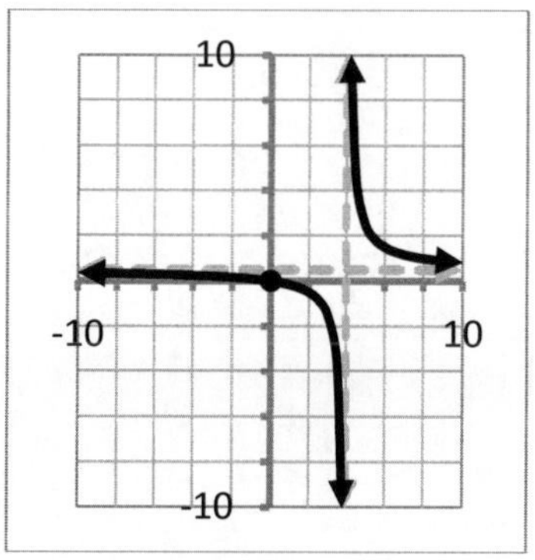

13. Graph the function $h(x) = 2^{x+1}$. Show its x- and y-intercepts if there are any.

a.
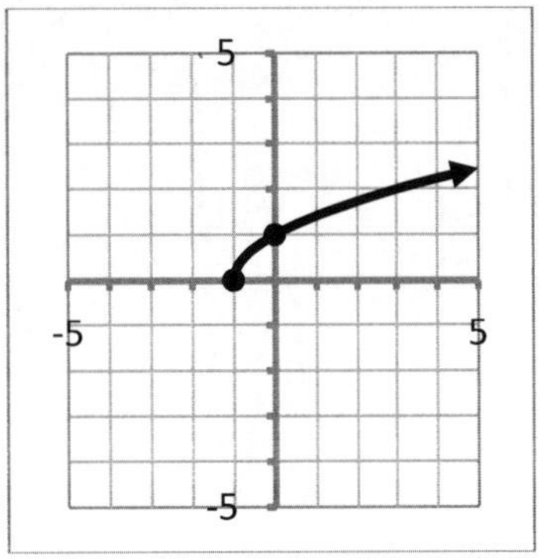

b.
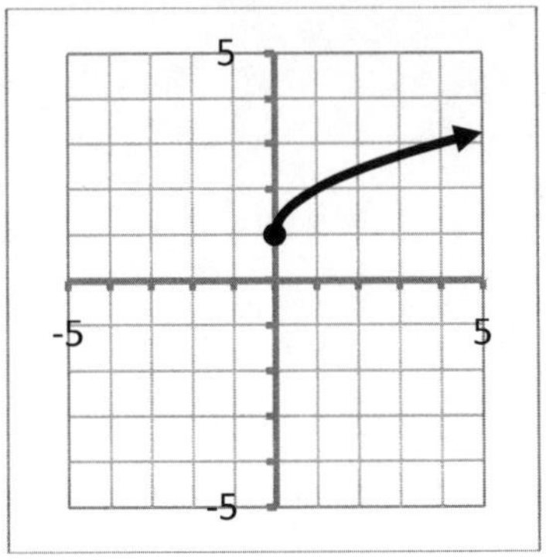

c.
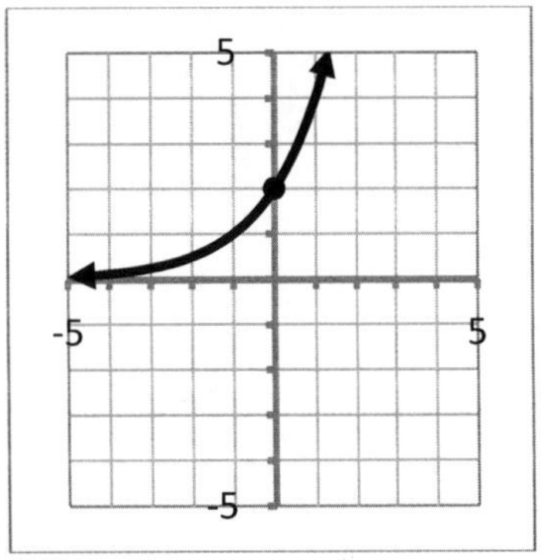

d.
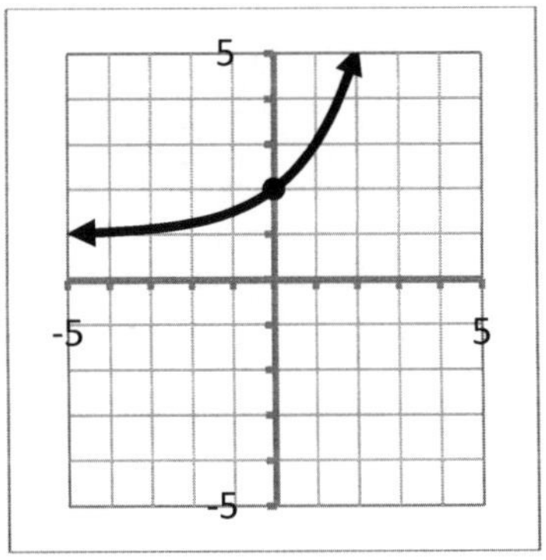

14. Graph the function $f(x) = \tan(-2x)$.

a.
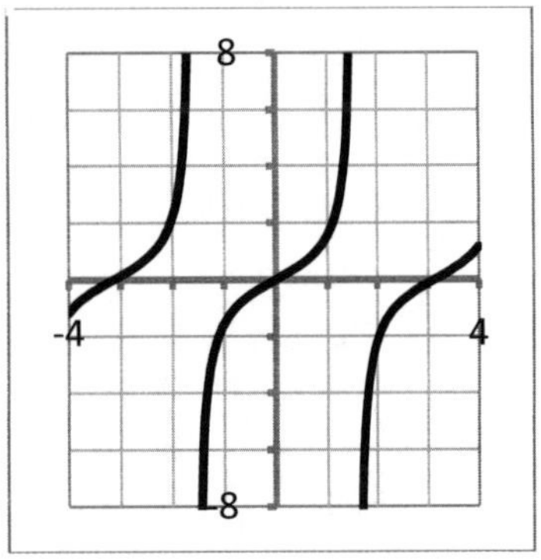

b.
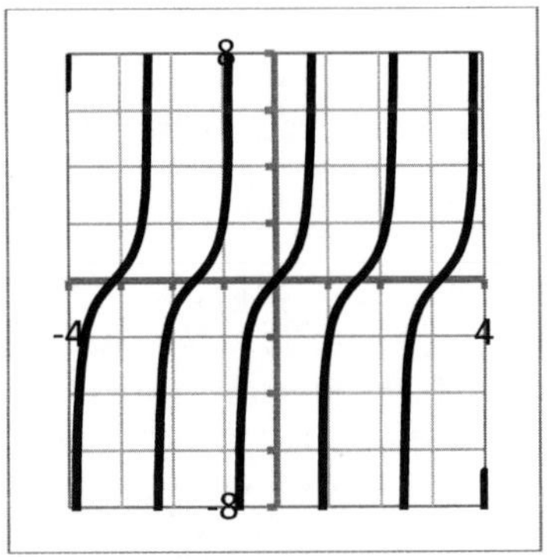

c.
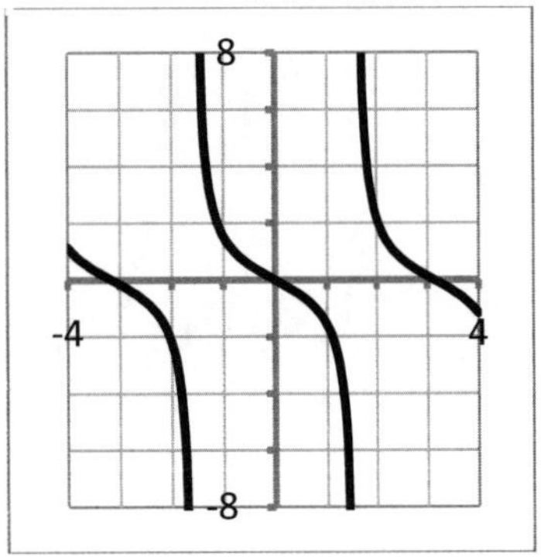

d.
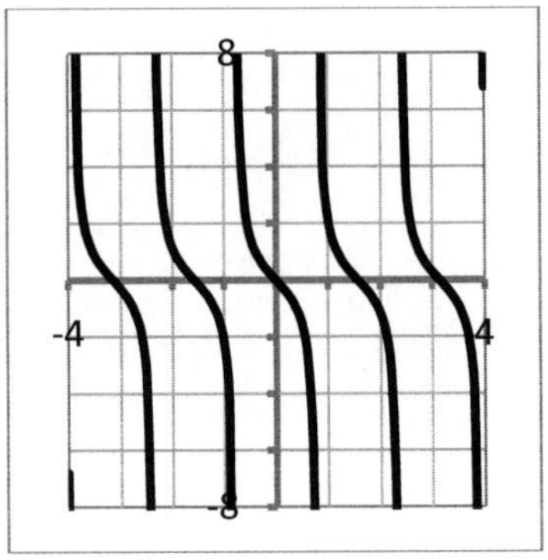

15. An object is thrown up in the air. Its height above the ground (in feet) t seconds after it is thrown is given by the function $h(t) = -16t^2 + 48t$. Use factoring to determine how long it will take (in seconds) for the object to hit the ground after it is thrown.

a. 3 seconds
b. 6 seconds
c. 8 seconds
d. 16 seconds

16. Which function represents exponential decay?

a. $y = (0.98)^t$
b. $y = (t - 0.56)^2$
c. $y = (1.05)^t$
d. $y = 100(1.03)^t$

17. Which function has the same y-intercept as the function graphed below?

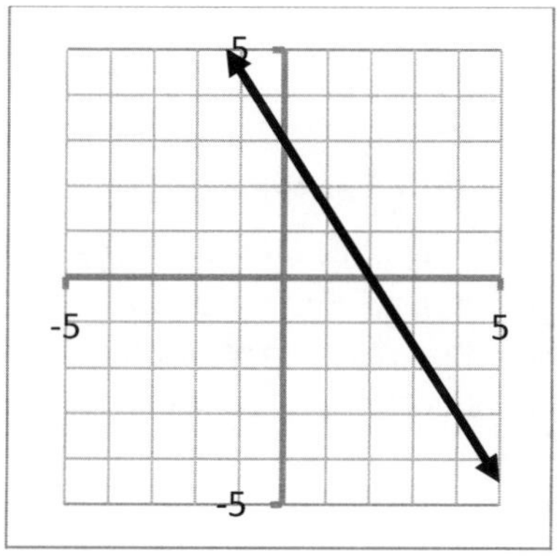

a. $g(x) = 2x - 4$
b. $g(x) = x^2 - 5x + 4$
c. $g(x) = x^3 + 3x^2 + x + 3$
d. $g(x) = \frac{4}{3x+2}$

18. Randi works as a salesperson in a high-end department store, where she is paid a 5% commission. In other words, she is paid 5% of the value of all sales made to her clients. Write a function $s(x)$ that represents her salary (in dollars) if the value of her clients' sales is x dollars.

a. $s(x) = 0.05x$
b. $s(x) = 5x$
c. $s(x) = \frac{x}{0.05}$
d. $s(x) = \frac{x}{5}$

19. A certain type of car is worth \$30,000 when you buy it. After that, it depreciates by about \$4,500 per year. Write a function $v(t)$ that represents the value of the car (in thousands of dollars) t years after you buy it.

a. $v(t) = 4.5t$
b. $v(t) = 30 - 4.5t$
c. $v(t) = 4{,}500t + 30{,}000$
d. $v(t) = 30{,}000 - 4{,}500t$

20. You are blowing up a balloon with air. The function $r(v) = \frac{1}{2}\sqrt[3]{v}$ represents the radius of the balloon (in cm) when it has a volume v of air in it. The function $s(r) = 12r^2$ represents the surface area of the balloon (in cm^2) when its radius is r cm. Write a function $s(v)$ for the surface area of the balloon when it has v cm^3 of air in it.

a. $s(v) = 3\sqrt[3]{v}$
b. $s(v) = 3v^{2/3}$
c. $s(v) = 6\sqrt[3]{v}$
d. $s(v) = 6v^{2/3}$

21. There are 100 bacteria in a Petri dish. The number of bacteria doubles every day, so that on the first day, there are 100 bacteria; on the second, there are 200; on the third, there are 400; and so on. Write a formula for the number of bacteria on the nth day.

a. $b(n) = 100 \times 2^{n-1}$
b. $b(n) = 100n$
c. $b(n) = 100n^2$
d. $b(n) = 200(n-1)$

22. Compare the graphs of $f(x) = \frac{x}{3x-6}$ and $g(x) = \frac{2x}{3x-6}$.

a. The graph of g is the graph of f shifted two units to the right.
b. The graph of g is the graph of f shifted two units up.
c. The graph of g is the graph of f stretched horizontally
d. The graph of g is the graph of f stretched vertically.

23. Compare the graphs of $f(x) = x^2 - 4$ and $g(x) = x^2 - 9$.

a. The graph of g is the graph of f shifted five units down.
b. The graph of g is the graph of f shifted five units up.
c. The graph of g is the graph of f shifted nine units down.
d. The graph of g is the graph of f shifted nine units up.

24. What is the inverse of the function $f(x) = \frac{5}{x+2}$ (where $x \neq -2$ and $x \neq 0$)?

a. $f^{-1}(x) = \frac{3}{x}$
b. $f^{-1}(x) = \frac{5}{x} - 2$
c. $f^{-1}(x) = \frac{x-2}{5}$
d. $f^{-1}(x) = \frac{x}{5} - 2$

25. What is the inverse of the function $f(x) = \sqrt[3]{x} - 6$?

a. $f^{-1}(x) = x^3 + 2$
b. $f^{-1}(x) = (x + 2)^3$
c. $f^{-1}(x) = x^3 + 6$
d. $f^{-1}(x) = (x + 6)^3$

26. The natural logarithm function $\ln x$ is the inverse of e^x. What is the composition of $\ln x$ and e^x?

a. $\ln(e^x) = 0$
b. $\ln(e^x) = 1$
c. $\ln(e^x) = \frac{1}{x}$
d. $\ln(e^x) = x$

27. The equation $y = g(x)$ is graphed below.

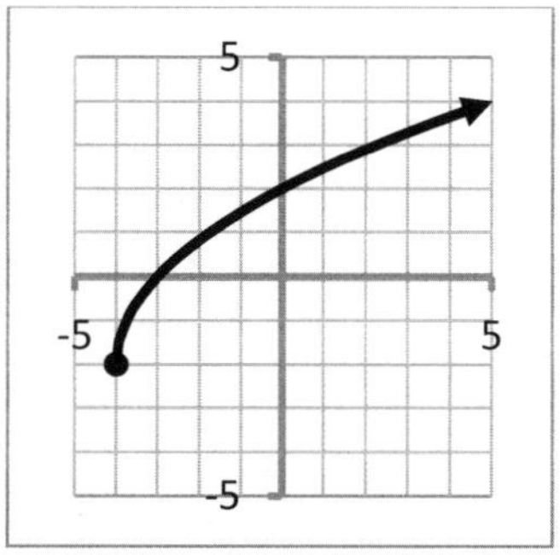

What is the value of $g^{-1}(0)$?

a. $g^{-1}(0) = -3$
b. $g^{-1}(0) = -2$
c. $g^{-1}(0) = 0$
d. $g^{-1}(0) = 2$

28. The table below displays the value of $h(x)$ for nine different values of x.

x	−4	−3	−2	−1	0	1	2	3	4
h(x)	−8	−6	−4	−2	8	4	2	1	0.5

What is the value of $h^{-1}(-4)$?

a. $h^{-1}(-4) = -8$
b. $h^{-1}(-4) = -2$
c. $h^{-1}(-4) = 1$
d. $h^{-1}(-4) = 0.5$

29. The function $f(x) = 3 - x^2$ is not invertible if the domain is all real numbers. Which domain makes the function invertible?

a. $x \geq -3$
b. $x \geq 0$
c. $x \leq \sqrt{3}$
d. $x \leq 3$

30. Solve the equation $\log_5(x^3 - 2) = 2$ for x.

a. $x = 2$
b. $x = 3$
c. $x = 8$
d. $x = 9$

31. Linear functions grow by equal differences over equal intervals. By what amount does the linear function $f(x) = 4x - 8$ grow by over every interval whose length is 5?

a. –40
b. 12
c. 20
d. 40

32. The amount of money a certain cell phone plan charges per month for a certain number of text messages can be modeled with a linear function. For example, if you make 40 text messages in a month, they charge $31; and if you make 60 text messages, they charge $34. Determine how much they charge if you make 80 text messages in a month.

a. $37
b. $38
c. $39
d. $40

33. The population of a certain city can be modeled with an exponential function. In 1960, its population was 4,000,000. In 1980, its population was 5,000,000. If this continued, what was the population of the city in the year 2000?

a. 6,000,000
b. 6,250,000
c. 6,500,000
d. 7,000,000

34. The linear function $g(x)$ is graphed below.

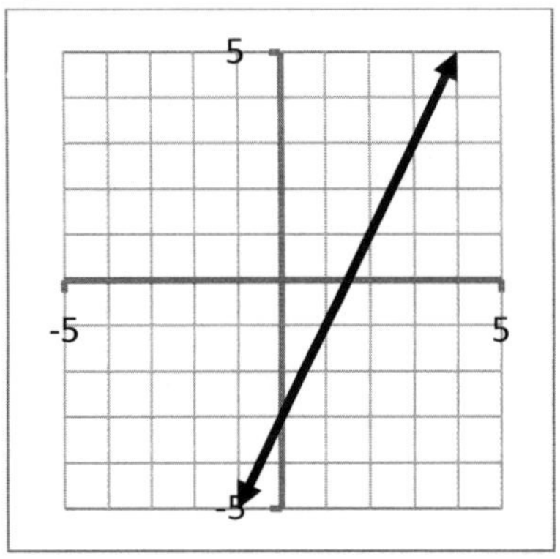

Write an explicit formula for $g(x)$.

a. $g(x) = -2x - 3$
b. $g(x) = -5x + 2$
c. $g(x) = 2x - 3$
d. $g(x) = 5x + 2$

35. The table below displays the value of the exponential function $h(x)$ for five different values of x.

x	−2	−1	0	1	2
h(x)	48	24	12	6	3

Write an explicit formula for $h(x)$.

a. $h(x) = 6(0.5)^x$
b. $h(x) = 6 \cdot 2^x$
c. $h(x) = 12(0.5)^x$
d. $h(x) = 12 \cdot 2^x$

36. Four functions are graphed below.

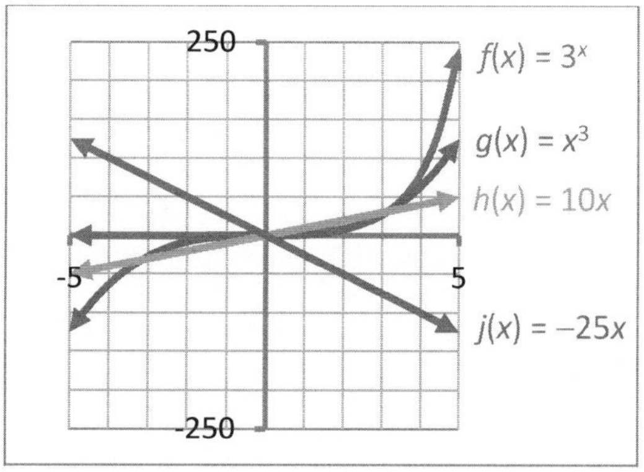

Which function eventually exceeds the other three as x increases?

a. $f(x) = 3^x$
b. $g(x) = x^3$
c. $h(x) = 10x$
d. $j(x) = -25x$

37. Solve the exponential equation $2 \cdot 4^{3t} = 18$ for t.

a. $t = \frac{\log_4 9}{3}$
b. $t = \log_4 3$
c. $t = \frac{\log_4 18}{3}$
d. $t = \frac{\log_4 18}{6}$

38. To rent a car, an agency charges a fixed fee plus an additional amount per day. The linear function $c(t) = 45t + 55$ represents the cost (in dollars) to rent a car for t days. After the fixed fee, how much does it cost to rent the car per day?

a. $10 per day
b. $45 per day
c. $55 per day
d. $100 per day

39. A certain isotope decays exponentially with a half-life of 1000 years. If you have a certain amount of this isotope, the remaining mass (in milligrams) after t millennia can be represented by the exponential function $m(t) = 10(0.5)^t$. How much of this isotope was present in the original sample?

a. 0.33 mg
b. 1 mg
c. 10 mg
d. 33 mg

40. In the figure below, circle O is a unit circle, and the length of $\widehat{AB}$ is $\frac{3\pi}{4}$.

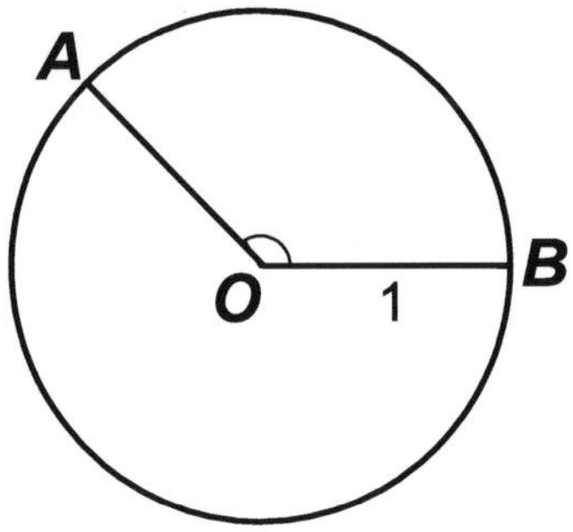

What is the measure of $\angle AOB$?

a. $\frac{\pi}{4}$
b. $\frac{3\pi}{4}$
c. $\frac{5\pi}{4}$
d. π

41. A unit circle and an angle are graphed on the coordinate plane below.

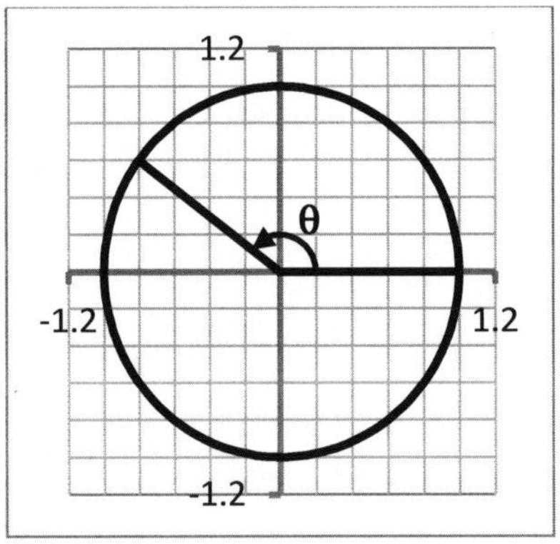

Use the graph to calculate the value of cos θ.

a. cos θ = –0.8
b. cos θ = –0.6
c. cos θ = 0.6
d. cos θ = 0.8

42. Calculate the value of $\sin\frac{\pi}{4}$.

a. $\sin\frac{\pi}{4} = \frac{\sqrt{2}}{2}$
b. $\sin\frac{\pi}{4} = \frac{\sqrt{3}}{2}$
c. $\sin\frac{\pi}{4} = 1$
d. $\sin\frac{\pi}{4} = \sqrt{2}$

43. Express $\cos(\pi + x)$ as a function of x.

a. $\cos(\pi + x) = -\cos x$
b. $\cos(\pi + x) = -\sin x$
c. $\cos(\pi + x) = \cos x$
d. $\cos(\pi + x) = \sin x$

44. Which of the following describes the properties of $\sin x$?

a. It is aperiodic and has even symmetry.
b. It is aperiodic and has odd symmetry.
c. It is periodic and has even symmetry.
d. It is periodic and has odd symmetry.

45. The voltage of an alternating current can be modeled with a cosine function. A certain current varies in voltage from +150 V to -150 V with a frequency of 60 Hz (cycles per second). Write a function $V(t)$ that represents the voltage (in volts) after t seconds.

a. $V(t) = \cos(60t) + 120$
b. $V(t) = \cos(120\pi t) + 120$
c. $V(t) = 150\cos(60t)$
d. $V(t) = 150\cos(120\pi t)$

46. The function $f(x) = \tan x$ is not invertible if the domain is all real numbers. Which domain makes the function invertible?

a. $-\pi < x < \pi$
b. $-\frac{\pi}{2} < x < \frac{\pi}{2}$
c. $0 \leq x \leq \frac{3\pi}{2}$
d. $0 \leq x \leq 2\pi$

47. A weight is suspended from a spring and released. The trigonometric function $h(t) = 20\cos t + 30$ represents the height of the weight above the ground (in inches) t seconds after it is released. Use an inverse function to determine when the weight is 15 inches above the ground for the first time. Round your answer to the nearest tenth of a second.

a. After about 0.7 seconds
b. After about 2.4 seconds
c. After about 3.9 seconds
d. After about 5.7 seconds

48. If $\cos\theta = \frac{12}{13}$ and θ is in quadrant IV, then what is $\sin\theta$?

a. $\sin\theta = -\frac{25}{169}$
b. $\sin\theta = -\frac{5}{13}$
c. $\sin\theta = \frac{25}{169}$
d. $\sin\theta = \frac{5}{13}$

49. Calculate $\cos\frac{7\pi}{12}$ using the addition formula for cosine.

a. $\cos\frac{7\pi}{12} = -\frac{1}{2}$
b. $\cos\frac{7\pi}{12} = \frac{\sqrt{2}-\sqrt{6}}{4}$
c. $\cos\frac{7\pi}{12} = \frac{\sqrt{6}-\sqrt{2}}{4}$
d. $\cos\frac{7\pi}{12} = \frac{1}{2}$

50. If $\tan x = \frac{2}{3}$ and $\tan y = \frac{1}{4}$, then what is $\tan(x + y)$?

a. $\tan(x + y) = \frac{5}{14}$
b. $\tan(x + y) = \frac{1}{2}$
c. $\tan(x + y) = \frac{11}{14}$
d. $\tan(x + y) = \frac{11}{10}$

Answers and Explanations

Math.Content.F.IF.1

1. B: The function f assigns to each element of the domain exactly one element of the range. Consequently, if x is any real number (including negative numbers), then the value of $f(x)$ is in the range, $f(x) \geq -2$. Therefore, the correct answer is choice B. On the other hand, choice A is incorrect because f is defined for all negative values of x, including ones less than -2. Choice C is incorrect because the range of f is $f(x) \geq -2$, not $f(-2)$. Finally, choice D is incorrect because the value of $f(-2)$ is defined since $x = -2$ is in the domain of f.

Math.Content.F.IF.2

2. B: To evaluate $g(4)$, substitute 4 for every occurrence of x in the equation $g(x) = 2x^2 - x + 3$. Then simplify the result using order of operations:

$$\begin{aligned} g(4) &= 2(4)^2 - (4) + 3 \\ &= 2(16) - 4 + 3 \\ &= 32 - 4 + 3 \\ &= 31 \end{aligned}$$

Math.Content.F.IF.2

3. C: The volume will be given by the expression $V(3)$. To calculate this value, substitute 3 for r in the equation $V(r) = \frac{4}{3}\pi r^3$. Then simplify the result using order of operations:

$$\begin{aligned} V(3) &= \frac{4}{3}\pi(3)^3 \\ &= \frac{4}{3}\pi \cdot 27 \\ &= 36\pi \end{aligned}$$

Therefore, the volume of the sphere is 36π.

Math.Content.F.IF.3

4. C: Since the function is defined recursively, you need to calculate all of the first six terms. The first and second terms are already given in the problem as $f(0) = 1$ and $f(1) = 1$. To calculate the third term, $f(2)$, write 2 as $(n + 1)$, or substitute 1 for n into the equation $f(n + 1) = f(n) + 2 \cdot f(n - 1)$:

$$\begin{aligned} f(1 + 1) &= f(1) + 2f(1 - 1) \\ f(2) &= f(1) + 2f(0) \end{aligned}$$

Since $f(0) = 1$ and $f(1) = 1$, substitute 1 for $f(0)$ and 1 for $f(1)$, and then simplify the result:

$$\begin{aligned} f(2) &= (1) + 2(1) \\ &= 3 \end{aligned}$$

Thus, the third term is 3. To calculate the fourth term, $f(3)$, write 3 as $(n + 1)$, or substitute 2 for n, 1 for $f(1)$, and 3 for $f(2)$ into the equation $f(n + 1) = 2 \cdot f(n) + f(n - 1)$ and simplify the result:

$$\begin{aligned} f(2 + 1) &= f(2) + 2f(2 - 1) \\ f(3) &= f(2) + 2f(1) \end{aligned}$$

$$= 3 + 2(1)$$
$$= 5$$

Continuing in this way, you will find that the first five terms of the sequence are 1, 1, 3, 5, 11, 21. Therefore, the sixth term of the sequence is 21.

Math.Content.F.IF.4

5. B: The pendulum reaches its lowest point when h is at its minimum. From the table, you can tell that the minimum is 2. Since $h(t) = 2$ when $t = 0.6$, the pendulum reaches its lowest point 0.6 seconds after it is released.

Math.Content.F.IF.5

6. B: The domain of a function is the set of all possible input values that can be evaluated by a function. In the given function, the input is t, which represents the year (e.g. $t = 1900$ represents the year 1900). Since the function represents the world population from 1900 to 2000, the domain should be $1900 \leq t \leq 2000$.

Math.Content.F.IF.6

7. D: The average rate of change of a function $f(x)$ from x_1 to x_2 is $\frac{f(x_2)-f(x_1)}{x_2-x_1}$. This formula is as analogous to the one for slope, rewritten in the context of functions. If $x_1 = 0$ and $x_2 = 6$, the average rate of the given function is given by expression below:

$$\frac{f(x_2) - f(x_1)}{x_2 - x_1} = \frac{f(6) - f(0)}{6 - 0}$$

Calculate the values of $f(0)$ and $f(6)$:

$f(0) = 4(0) - 7 = -7$

$f(6) = 4(6) - 7 = 17$

Then substitute these values into the expression for the average rate of change and simplify the result:

$$\frac{f(6) - f(0)}{6 - 0} = \frac{(17) - (-7)}{6 - 0}$$
$$= \frac{24}{6}$$
$$= 4$$

Math.Content.F.IF.7a

8. D: The graph of a linear function is a straight line. To graph the given function, first create a table of values.

x	f(x) = –3x + 3	(x,y)
0	–3(0) + 3 = 3	(0,3)
1	–3(1) + 3 = 0	(–3,0)
2	–3(2) + 3 = –3	(2,0)

Next plot the three points on a coordinate axis and connect them with a straight line. The *x*- and *y*-intercepts are the points where the line passes through the *x*- and *y*-axes, respectively.

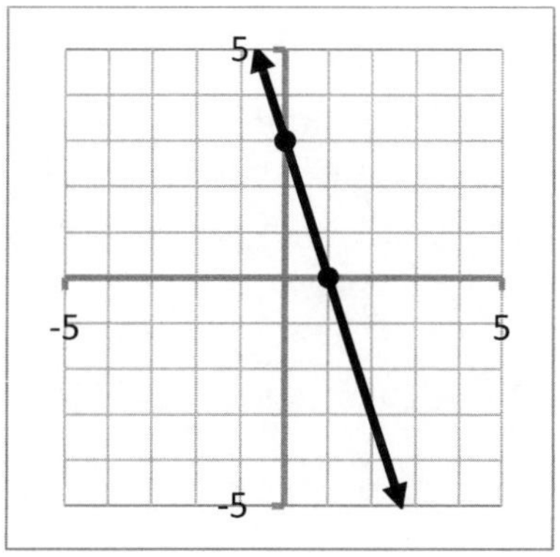

Math.Content.F.IF.7a

9. D: The graph of a quadratic function is a curve. To graph the given function, first create a table of values.

x	$g(x) = -x^2 + 4x - 3$	(x,y)
0	$-(0)^2 + 4(0) - 3 = -3$	(0,–3)
1	$-(1)^2 + 4(1) - 3 = 0$	(1,0)
2	$-(2)^2 + 4(2) - 3 = 1$	(2,1)
3	$-(3)^2 + 4(3) - 3 = 0$	(3,0)
4	$-(4)^2 + 4(4) - 3 = -3$	(4,–3)

Next plot the five points on a coordinate axis and connect them with a curve. The graph will be an upside-down u-shaped parabola. Plot the maximum, which is the highest point on the graph.

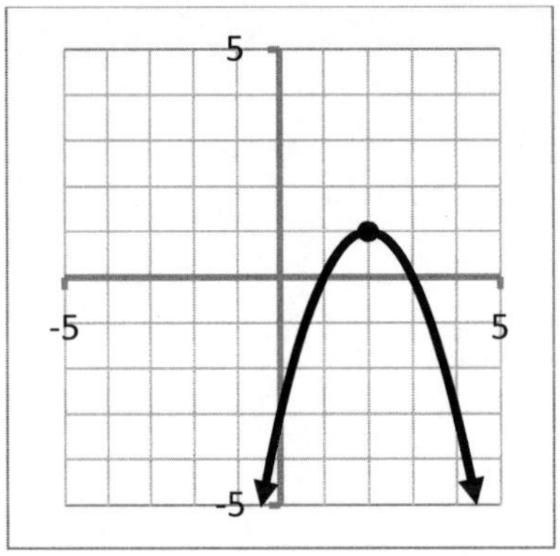

Math.Content.F.IF.7b

10. A: The absolute value of a number (represented by vertical lines around the number) makes the number positive if it is negative and keeps it the same if it is positive or zero. To graph the given function, first create a table of values.

x	h(x) = |x + 2| – 1	(x,y)
–4	|(–4) + 2| – 1 = 1	(–4,1)
–3	|(–3) + 2| – 1 = 0	(–3,0)
–2	|(–2) + 2| – 1 = –1	(–2, –1)
–1	|(–1) + 2| – 1 = 0	(–1,0)
0	|(0) + 2| – 1 = 1	(0,1)

Next plot the five points on a coordinate axis and connect them with two straight lines. The graph will be a V-shaped graph.

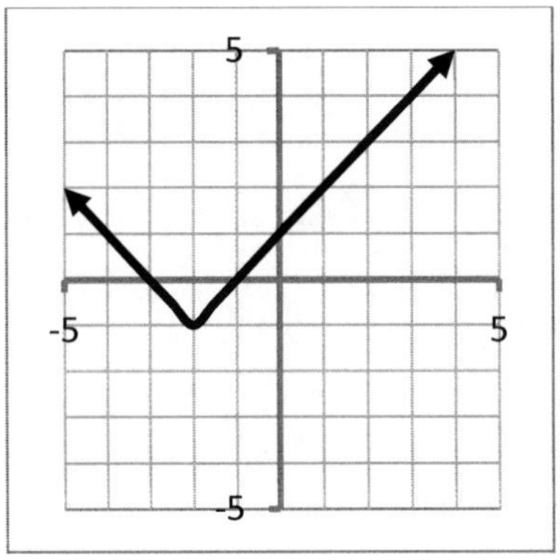

Math.Content.F.IF.7c

11. C: Graph the given function by either creating a table of values by hand or by inputting the function $y = x^3 - 2x^2 - 4x + 8$ into a graphing calculator.

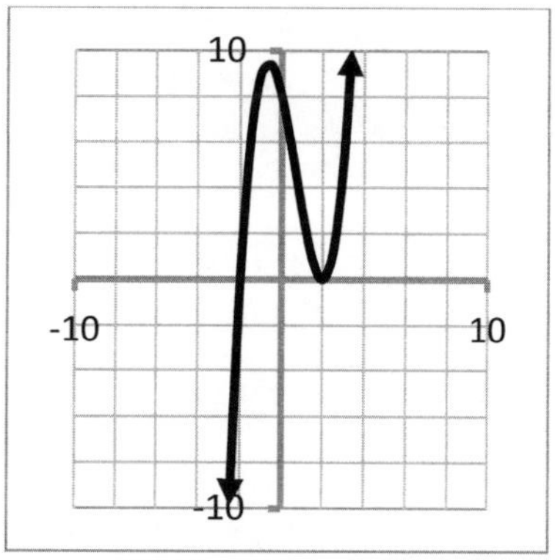

The zeroes of a function are the values where the function equals zero, so in this case, where $f(x) = 0$. Graphically, the zeroes occur where the graph crosses the x-axis. Notice from the graph above that f crosses the x-axis two times: at the points $(-2,0)$, and $(2,0)$. Plot these points on the graph of f.

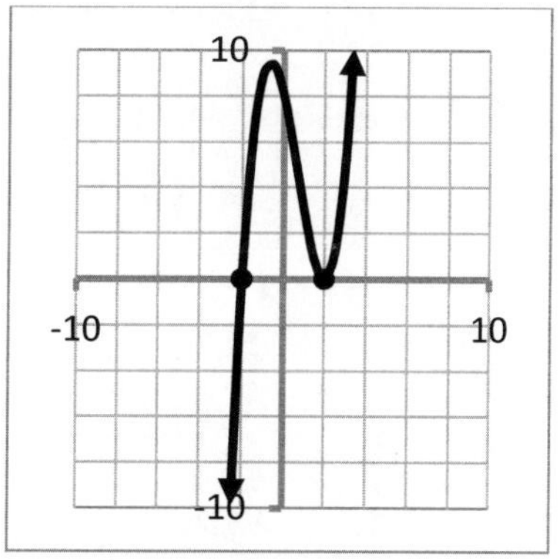

Math.Content.F.IF.7d

12. A: Graph the given function by either creating a table of values by hand or by inputting the function $y = \frac{3x}{2x^2-8}$ into a graphing calculator.

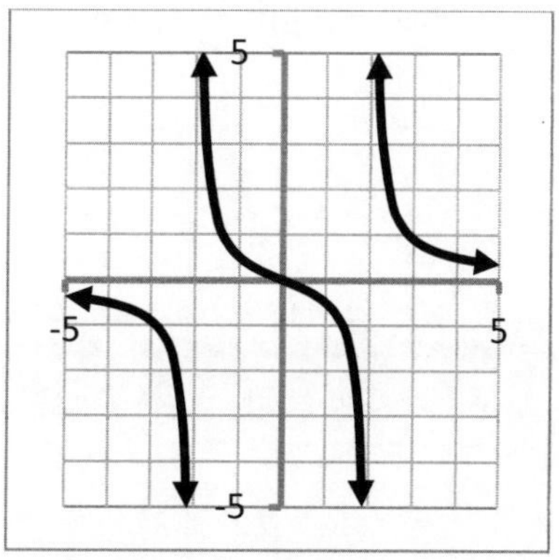

The zeroes of a function are the values where the function equals zero, so in this case, where $g(x) = 0$. You might be able to tell from your table of values this occurs at when $x = 0$. Graphically, the zeroes occur where the graph crosses the x-axis. Notice from the graph above, that g crosses the x-axis at the origin (0,0). Plot this point on the graph of g.

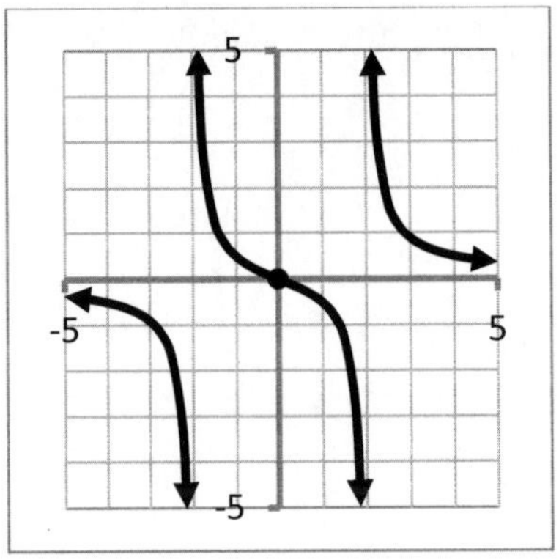

An asymptote is a line that the graph of a function gets closer and closer to but never touches. Notice that the function never touches the vertical lines $x = -2$ and $x = 2$. (Check: if $x = \pm 2$, $g(\pm 2) = \frac{3(\pm 2)}{2(\pm 2)^2 - 8} = \pm \frac{6}{8-8}$, whfich is undefined! There can be no value for the function at these values of x.)Therefore, $x = -2$ and $x = 2$ are both vertical asymptotes of f. Graph these lines on the same graph with light dotted lines.

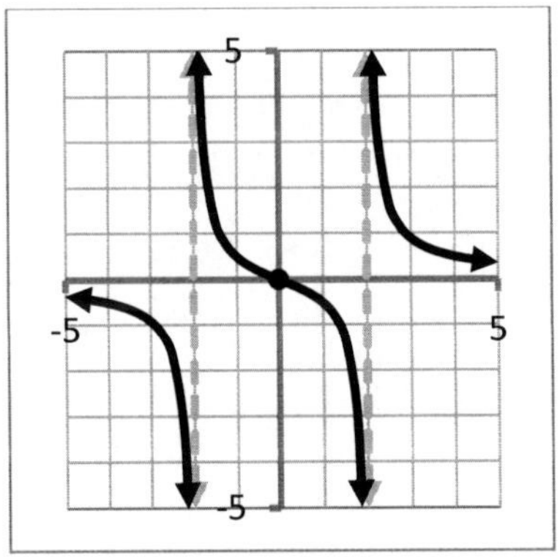

Math.Content.F.IF.7e

13. C: Graph the given function by either creating a table of values by hand or by inputting the function $y = 2^{x+1}$ into a graphing calculator. In addition, plot the points where the line passes through the x- and y-axes, which are the x- and y-intercepts of h.

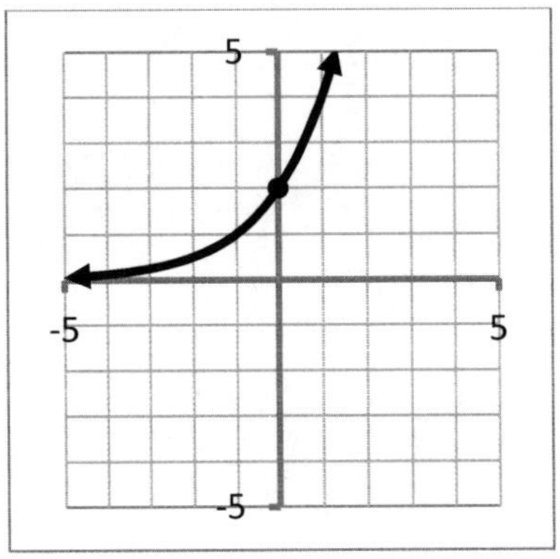

Math.Content.F.IF.7e

14. D: Graph the given function by either creating a table of values by hand or by inputting the function $y = \tan(-2x)$ into a graphing calculator. The graph of a trigonometric function like tangent is a periodic function. That is, it repeats its values after a certain interval. The period of a tangent function is given by π/b, where b is the coefficient of x in the argument of the tangent function; in this case, the period is $\pi/2$. Verifying that tan(-2x) is positive or negative for a particular value of x, say $\pi/6$ or 30° (in which case $f(x)$ is negative), determines whether the answer is B or D.

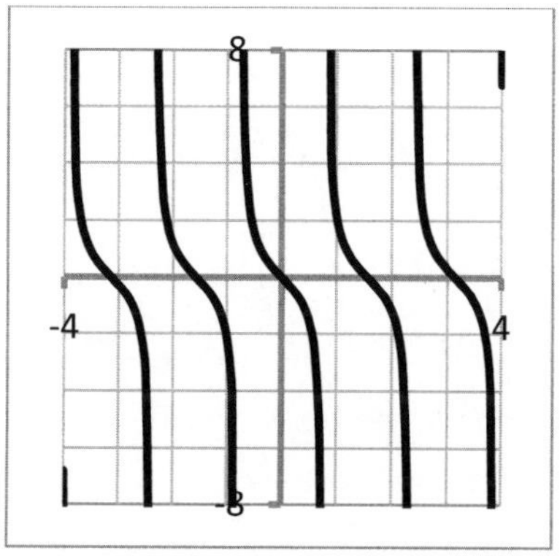

Math.Content.F.IF.8a

15. A: The object will hit the ground when its height is zero. In mathematical notation, this will happen when $h(t) = 0$. Set up an equation when $h(t) = 0$ and then solve it for t. The first step is to factor the quadratic expression:

$$\begin{aligned} h(t) &= 0 \\ -16t^2 + 48t &= 0 \\ -16t(t-3) &= 0 \end{aligned}$$

Next, use the zero-product rule to solve the result. Using the factorization of h, write two equations and solve them both for x:

$$-16t = 0 \text{ or } t - 3 = 0$$

$$t = 0\, t = 3$$

The answer $t = 0$ does not make sense (since that corresponds to when it is thrown), so we can discard that value. Thus, the object will hit the ground exactly 3 seconds after it is thrown.

Math.Content.F.IF.8b

16. A: An exponential function can be written in the form $y = a \cdot b^{t/n}$. In this form, if $b > 1$, then the function represents exponential growth (that is, its value increases as t gets bigger). And if $0 < b < 1$, then the function represents exponential decay (that is, its value decreases as t gets bigger). Examining the given choices, the only function that represents exponential decay is $y = (0.98)^t$. To verify this, plug in a few increasing values for t into each equation: the only one for which $y(t)$ decreases is A.

Math.Content.F.IF.9

17. C: The y-intercept of a function is the point where the line passes through the y-axis; it therefore has an x-coordinate of zero. Since the graphed function passes through the y-axis at $(0,3)$, its y-intercept is 3. To determine which function has the same y-intercept, substitute 0 for x into each function and see if the result is 3. Start with the choice A, which is $g(x) = 2x - 4$.

$$2(0) - 4 = 0 - 4$$

$$= -4$$

Since the result is not 3, the y-intercept is not 3. Next try choice B, $g(x) = x^2 - 5x + 4$.

$$(0)^2 - 5(0) + 4 = 0 - 0 + 4$$

$$= 4$$

The y-intercept of the second function is not 3 either. Try choice C, $g(x) = x^3 + 3x^2 + x + 3$.

$$\begin{aligned}(0)^3 + 3(0)^2 + (0) + 3 &= 0 + 0 + 0 + 3 \\ &= 3\end{aligned}$$

Thus, the function $g(x) = x^3 + 3x^2 + x + 3$ has the same y-intercept as the graphed function.

Math.Content.F.BF.1a

18. A: Since Randi's commission is 5% of her sales, you can calculate her salary by multiplying her clients' sales by 5%. Since 5% is equivalent to 0.05, the correct function is $s(x) = 0.05x$.

Math.Content.F.BF.1b

19. B: The value of the car is its initial value (i.e. when it's new) minus the amount that it has depreciated. We are told that the car is worth \$30,000 when it is new. Therefore, since v is in thousands of dollars, the initial value is given by the constant function $v_1(t) = 30$. Furthermore, because the value decreases by \$4,500 per year, the amount it depreciates by is given by the linear function $v_2(t) = 4.5t$. Finally, to find the value of the car after t years, subtract the two functions.

$$\begin{aligned}v(t) &= v_1(t) - v_2(t) \\ &= 30 - 4.5t\end{aligned}$$

Math.Content.F.BF.1c

20. B: The first function $r(v)$ gives the radius of the balloon after being filled by v cm^3 of air. The second function $s(r)$ gives the surface area of the balloon when its radius is r cm. The result of the first is the argument of the second. Consequently, the composition of the functions $s(r(v))$ surface area of the balloon when it has v cm^3 of air. Calculate $s(r(v))$.

$$\begin{aligned}s\big(r(v)\big) &= 12[r(v)]^2 \\ &= 12\left[\frac{1}{2}\sqrt[3]{v}\right]^2 \\ &= 12\left[\frac{1}{4}v^{2/3}\right] \\ &= 3v^{2/3}\end{aligned}$$

Thus, the function $s(v) = 3v^{2/3}$ represents the surface area of the balloon when it has v cm^3 of air.

Math.Content.F.BF.2

21. A: The number of bacteria forms a geometric sequence: 100, 200, 400, 800, 1,600, etc. Notice that, if you ignore the two zeroes, these numbers are all powers of 2 (i.e. $2^0, 2^1, 2^2, 2^3, 2^4$, etc.). In other words, they are all 100 multiplied by a power of 2, so the sequence can be written as an exponential function of the form $y = 100 \times 2^n$. However, since this sequence begins with an exponent of 0 on the first day, rather than 1, and an exponent of 1 on the second day rather than 2, you need to subtract 1 from n to get the correct power. Thus, the function $b(n) = 100 \times 2^{n-1}$ represents the number of bacteria on the nth day.

Math.Content.F.BF.3

22. D: To transform $f(x)$ into $g(x)$, you have to multiply it by 2. This transformation stretches the graph vertically, making it two times bigger vertically.

Math.Content.F.BF.3

23. A: To transform $f(x)$ into $g(x)$, you have to subtract 5 from f. This transformation results in a translation, or shift, of the graph. In particular, it shifts the graph five units down.

Math.Content.F.BF.4a

24. B: The inverse of a function is another function that *undoes* the original one. By convention, the inverse of $f(x)$ is written as $f^{-1}(x)$. To find the inverse of $f(x)$, first replace $f(x)$ with y. Then switch all the x's with all the y's (and vice versa):

$f(x) = \frac{5}{x+2}$ (original)

$y = \frac{5}{x+2}$ (replace $f(x)$ with y)

$x = \frac{5}{y+2}$ (exchange x and y)

Next solve the equation for y:

$$x(y + 2) = 5$$
$$y + 2 = \frac{5}{x}$$
$$y = \frac{5}{x} - 2$$

Finally, replace y with $f^{-1}(x)$ to get the inverse.

$$f^{-1}(x) = \frac{5}{x} - 2$$

You can see now why the restriction $x \neq 0$ is required: if $x = 0$, this inverse function is undefined!

Math.Content.F.BF.4a

25. D: The inverse of a function is another function that *undoes* the original one. By convention, the inverse of $f(x)$ is written as $f^{-1}(x)$. To find the inverse of $f(x)$, first replace $f(x)$ with y. Then switch all the x's with all the y's (and vice versa):

$$f(x) = \sqrt[3]{x} - 6 \qquad \text{(original)}$$
$$y = \sqrt[3]{x} - 6 \qquad \text{(replace } f(x) \text{ with y)}$$
$$x = \sqrt[3]{y} - 6 \qquad \text{(exchange x and y)}$$

Next solve the equation for y:

$$x + 6 = \sqrt[3]{y}$$
$$(x + 6)^3 = \left(\sqrt[3]{y}\right)^3$$
$$(x + 6)^3 = y$$

Finally, replace y with $f^{-1}(x)$ to get the inverse.

$$f^{-1}(x) = (x + 6)^3$$

Math.Content.F.BF.4b

26. D: The inverse of a function is another function that *undoes* the original one. For a function f^{-1} to be the inverse of f, their composition must be equal to x as shown below:

$$f^{-1}(f(x)) = x$$

$$f(f^{-1}(x)) = x$$

Therefore, the composition of the given functions, $\ln x$ and e^x, is the variable x.

$$\ln(e^x) = x$$

Math.Content.F.BF.4c

27. A: The inverse of a function is another function that *undoes* the original one. By convention, the inverse of $g(x)$ is written as $g^{-1}(x)$. For any value of x, the result of inputting x into g can itself be inputted into g^{-1}, and it will return the original value x. In other words, if $g(x) = y$, then $g^{-1}(y) = x$.

In regard to the problem, this means that the value of $g^{-1}(0)$ is the value of x for which $g(x) = 0$. Examining the graph of g, notice that it passes through the point $(-3,0)$. Therefore, $g(-3) = 0$, and the value of $g^{-1}(0)$ is -3.

Math.Content.F.BF.4c

28. B: The inverse of a function is another function that *undoes* the original one. By convention, the inverse of $h(x)$ is written as $h^{-1}(x)$. For any value of x, the result of inputting x into h can itself be inputted into h^{-1}, and it will return the original value x. In other words, if $h(x) = y$, then $h^{-1}(y) = x$.

In regard to the problem, this means that the value of $h^{-1}(-4)$ is the value of x for which $h(x) = -4$. Examining the table for h, notice that when $x = -2$, $h(x) = -4$. Therefore, $h(-2) = -4$, and the value of $h^{-1}(-4)$ is -2.

Math.Content.F.BF.4d

29. B: A function is invertible if its inverse is a function. The given function *f* is not invertible if the domain is all real numbers because in some cases different values of *x* result in the same value of *f*(*x*). Graphically, this means that the function does not pass the horizontal line test (which means that after reversing x and y the inverse does not pass the vertical line test). That is, at least one horizontal line (like $y = -1$) intersects the graph at more than one point (the points (–2, –1) and (2, –1)):

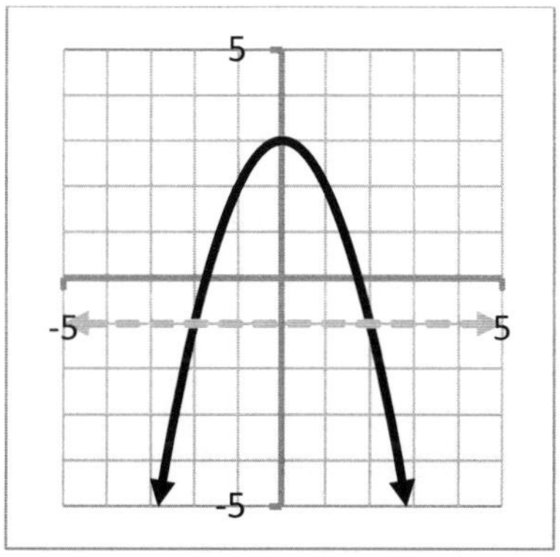

To make the function invertible, restrict the domain so that the new function does pass the horizontal line test. The only answer choice that accomplishes this is B: if you restrict the domain to $x \geq 0$, the resulting function will pass the horizontal line test as shown in the graph below.

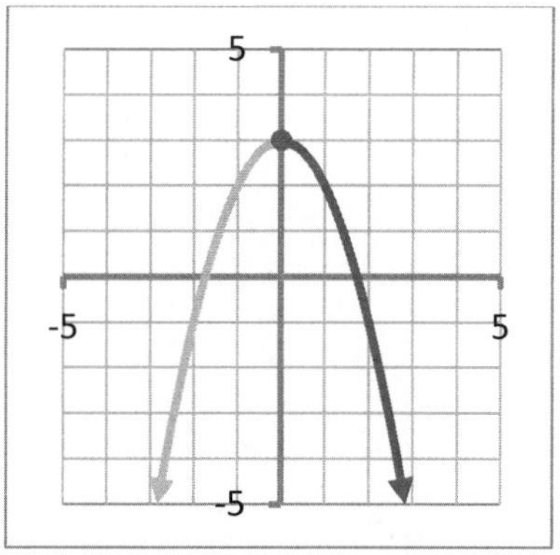

Math.Content.F.BF.5

30. B: The logarithm of a number is the exponent that the base must to be raised to in order to get that number. For example, since $2^4 = 16$, it is also true that $\log_2 16 = 4$. By this definition, the given equation is equivalent to the following equation:

$$x^3 - 2 = 5^2$$

Simplify this exponential equation and solve for *x*.

$$\begin{aligned} x^3 - 2 &= 25 \\ x^3 &= 27 \\ x &= 3 \end{aligned}$$

Thus, the solution is $x = 3$.

Math.Content.F.LE.1a

31. C: The length of an interval is the difference between its endpoints. For example, the length of the interval [2, 4] is 2. To determine how the given function grows over an interval of length 5, determine the value of *f* at each endpoint of that interval. Since linear functions grow by equal differences over equal intervals, you can use any interval of length 5, and your answer will apply to all such intervals. For example, you can use the interval [0, 5]:

$$\begin{aligned} f(0) &= 4(0) - 8 \\ &= 0 - 8 \\ &= -8 \\ f(5) &= 4(5) - 8 \\ &= 20 - 8 \\ &= 12 \end{aligned}$$

Since $f(0) = -8$ and $f(3) = 12$, the function increases by $12 - (-8) = 20$ over this interval, and over any interval of length 5.

Math.Content.F.LE.1b

32. A: Linear functions grow by equal differences over equal intervals. In other words, if the linear function $c(m)$ represents the monthly cost (in dollars) if you make m text messages, then intervals of equal length (in m) result in equal increases in the value of the function $c(m)$. From the problem, we know that $c(40) = 31$ and $c(60) = 34$. Thus, we can conclude that intervals of length 20 (like the interval [40, 60]) result in an increase of 3 since $34 - 31 = 3$. Therefore, since the length of [60, 80] is also 20, the function $c(m)$ increases by 3 over this interval as well. Use this information to calculate $c(80)$:

$$\begin{aligned} c(80) &= c(60) + 3 \\ &= 34 + 3 \\ &= 37 \end{aligned}$$

Therefore, we can conclude that the plan will charge \$37 if you make 80 text messages in a month.

Math.Content.F.LE.1c

33. B: Exponential functions grow by equal factors over equal intervals. In other words, if the exponential function $p(y)$ represents the population of the city in the year y, then intervals of equal length (in y) multiply the value of the function $p(y)$ by equal factors. From the problem, we know that $p(1960) = 4{,}000{,}000$, and $p(1980) = 5{,}000{,}000$. Use division to determine how intervals of length 20 (like the interval [1960, 1980]) change the value of $p(y)$:

$$\begin{aligned} \frac{p(1980)}{p(1960)} &= \frac{5{,}000{,}000}{4{,}000{,}000} \\ &= 1.25 \end{aligned}$$

Therefore, increasing t by 20 multiplies the value of $p(y)$ by 1.25. Therefore, since the length of [1980, 2000] is also 20, the value of $p(y)$ is multiplied by 1.25 over this interval as well. Use this information to calculate $p(2000)$:

$$\begin{aligned} p(2000) &= p(1980) \cdot 1.25 \\ &= 5{,}000{,}000 \cdot 1.25 \end{aligned}$$

$$= 6{,}250{,}000$$

Therefore, we can conclude that there were 6,250,000 people in the city in the year 2000.

Math.Content.F.LE.2

34. C: Linear functions can be written in the form $g(x) = ax + b$. To determine the value of a, notice that $g(x)$ increases by 2 every time x increases by 1. For instance, when x goes from 0 to 1, the value of y goes from –3 to –1; and when x goes from 1 to 2, the value of y increases from –1 to 1. Since $g(x)$ increases by 2, the value of a is 2, so the function is $g(x) = 2x + b$. Next calculate the value of b. From the graph, you can tell that $g(0) = –3$. Use this to calculate b:

$$\begin{aligned} g(0) &= -3 \\ 2(0) + b &= -3 \\ 0 + b &= -3 \\ b &= -3 \end{aligned}$$

Thus, the function is $g(x) = 2x – 3$.

Alternatively, writing an explicit formula in the form $g(x) = ax +$ b means solving for two variables, a and b. Use the information given in the graph to write two equations relating these two variables in terms of x and $g(x)$:

$$\begin{aligned} -3 &= a(0) + b \\ 1 &= a(2) + b \end{aligned}$$

The first readily solves to $b = -3$, which you can substitute into the second to solve for a:

$$1 = a(2) - 3$$
$$4 = a(2) = 2a$$
$$a = 2$$

Thus, the function is $g(x) = 2x - 3$.

Math.Content.F.LE.2

35. C: Exponential functions can be written in the form $h(x) = ab^x$. To determine the value of b, notice that $h(x)$ is halved every time x increases by 1. For instance, when x goes from –2 to –1, the value of the function drops (by half) from 48 to 24; and when x goes from –1 to 0, the function drops again from 24 to 12. Since the function is halved for every increase in x, the value of b is 0.5, so the function is $h(x) = a(0.5)^x$.

Next calculate the value of a. From the table, you can tell that $h(0) = 12$. Use this to calculate a.

$$\begin{aligned} h(0) &= 12 \\ a(0.5)^0 &= 12 \\ a(1) &= 12 \\ a &= 12 \end{aligned}$$

Thus, the function is $h(x) = 12(0.5)^x$.

Math.Content.F.LE.3

36. A: Examine the graph to determine which graph eventually exceeds the other three. In other words, find the function that eventually goes above the other three as the graph goes from left to right. Notice that the function *f* goes above the other three when $x = 5$.

Math.Content.F.LE.4

37. A: First divide both sides of the equation by 2:

$$2 \cdot 4^{3t} = 18$$
$$4^{3t} = 9$$

Next, take the logarithm of both sides to get rid of the exponent. Since the base of the exponent is 4, use 4 for the base of the logarithm as well so that you can simplify the result using the fact that $\log_4 4^x = x$ by definition:

$$\log_4 4^{3t} = \log_4 9$$
$$3t = \log_4 9$$

Finally, divide both sides by 3 to solve for *t*:

$$t = \frac{\log_4 9}{3}$$

Math.Content.F.LE.5

38. B: The linear function $c(t) = 45h + 55$ starts at 55 when $t = 0$ and increases by 45 every time *t* increases by 1. Therefore, the additional cost is $45 per day.

Math.Content.F.LE.5

39. C: The exponential function $m(t) = 10(0.5)^t$ starts at 10 when $h = 0$ and is multiplied by 0.5 every time *t* increases by 1 millennium. Therefore, you started with 10 mg of the isotope.

Math.Content.F.TF.1

40. B: An arc is a piece of a circle. In the figure, $\widehat{AB}$ is the piece of the circle that starts at point *A* and ends at *B*. In a unit circle, the length of an arc is the measure of the angle (in radians) subtended by the angle made by radii drawn to the endpoints of the arc. Therefore, the measure of $\angle AOB$ is equal to the arc length of $\widehat{AB}$, so its measure is $\frac{3\pi}{4}$.

Math.Content.F.TF.2

41. A: In a unit circle, trigonometric functions can be represented by considering angles that start from the positive side of the *x*-axis and are measured counterclockwise around the circle. For angles like this, the cosine of the angle is the *x*-coordinate of the point where the angle's side meets the unit circle. Since one side of the angle meets the unit circle at $(-0.8, 0.6)$, the value of $\cos\theta$ is -0.8.

Math.Content.F.TF.3

42. A: The angle $\pi/4$ is in radians. To convert it to degrees, multiply by $\frac{180}{\pi}$:

$$\frac{\pi}{4} \cdot \frac{180}{\pi} = 45°$$

Therefore, $\sin\frac{\pi}{4} = \sin 45°$. To calculate this value, draw a 45-45-90 triangle, which is a special triangle whose proportions you may have memorized. Making the hypotenuse one unit long simplifies things, although it is not necessary as long as the proportions are the same.

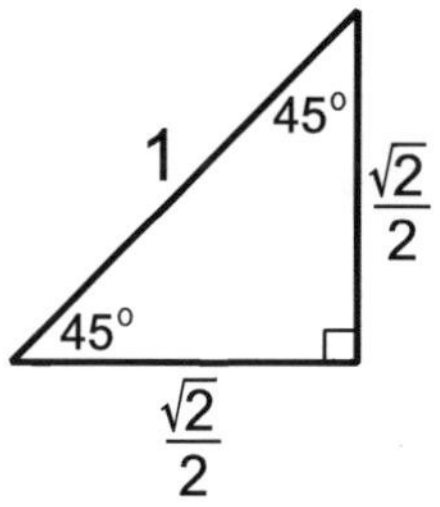

By SOH-CAH-TOA, the sine function is defined as $\frac{\text{opposite}}{\text{hypotenuse}}$ in a right triangle. Therefore, the value of $\sin\frac{\pi}{4}$ is $\frac{\sqrt{2}/2}{1}$, or $\frac{\sqrt{2}}{2}$.

Math.Content.F.TF.3

43. A: All trigonometric functions can be visualized with a unit circle on a coordinate plane. If an angle is measured from the positive side of the *x*-axis in the counterclockwise direction, then the cosine function is the *x*-coordinate of the point where the angle's other side meets the unit circle. To visualize the problem, draw a unit circle on a coordinate plane showing an arbitrary angle *x* and the angle π + *x*.

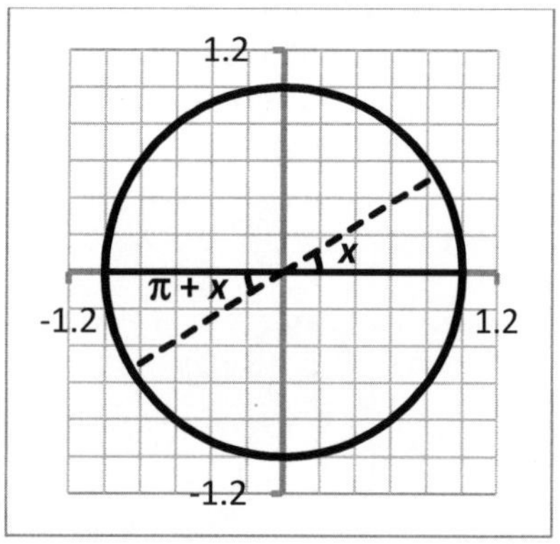

Notice that the *x*-coordinate associated with the angle *x* is the opposite (the negative value) of the *x*-coordinate associated with π + *x*. Therefore, cosine of *x* and cosine of π + *x* are opposites, so we can conclude that cos(π + *x*) = −cos *x*.

Math.Content.F.TF.4

44. D: Graph the trigonometric function $y = \sin x$.

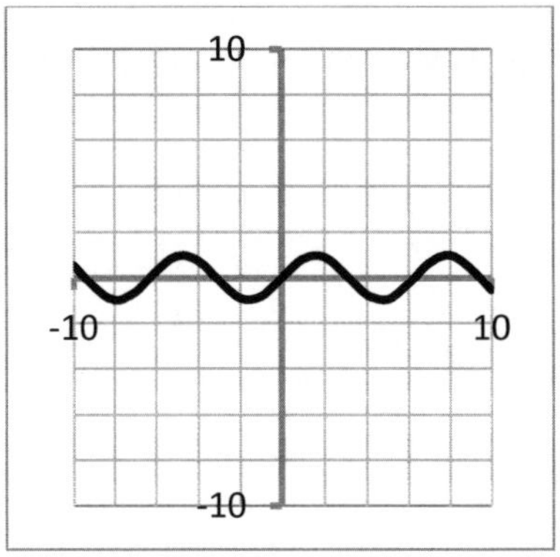

A function is periodic if it repeats its value in regular intervals and aperiodic if it doesn't. Since the graph of $y = \sin x$ repeats itself it is periodic. A function has even symmetry if it is symmetric about the y-axis and odd symmetry if it can be rotated 180° about the origin and look the same. Therefore, the graph has odd symmetry. Furthermore, you might see that for a given angle, say 45°, replacing it with -45° negates the y-component of the point of intersection with the unit circle, so that $\sin(-x) = -\sin(x)$, which is odd symmetry.

Math.Content.F.TF.5

45. D: The frequency of a periodic function is the number of cycles of the function per unit time. Since the voltage repeats 60 times per second, the frequency of $V(t)$ is 60. The amplitude is half the distance between the lowest and the highest point: $A = ½[150, -150] = 150$. The midline is given by the lowest point plus the amplitude or the highest minus the amplitude: $-150 + 150 = 150 - 150 = 0$.

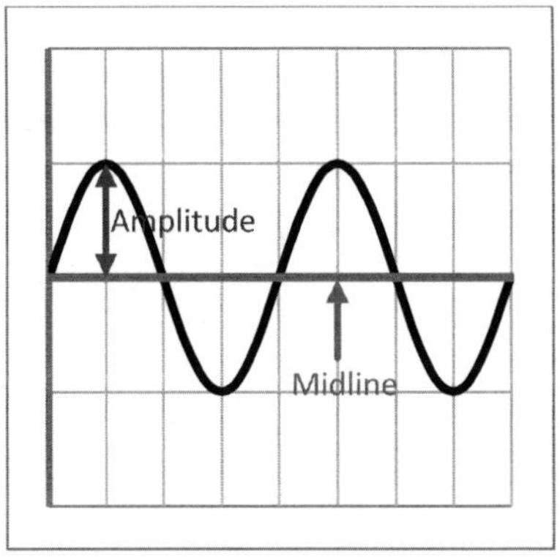

Now write the function $V(t)$ by modifying the parent function $y = \cos t$. The midline of $\cos t$ is 0 (so no change is required), but its amplitude is 1. To make the amplitude 150, multiply the parent function by 150.

$$y = 150\cos t$$

The frequency of this function is $\frac{1}{2\pi}$ since t advances by 2π units before the function repeats again. To make the frequency 60, multiply t by 120π so it repeats 60 times per second:

$$V(t) = 150\cos(120\pi t)$$

Math.Content.F.TF.6

46. B: A function is invertible if its inverse is a function. The given function *f* is not invertible if the domain is all real numbers because in some cases different values of *x* result in the same value of *f*(*x*). Graphically, this means that the function does not pass the horizontal line test (which means that after exchanging x and y the inverse does not pass the vertical line test). That is, at least one horizontal line (like $y = 2$) intersects the graph at more than one point:

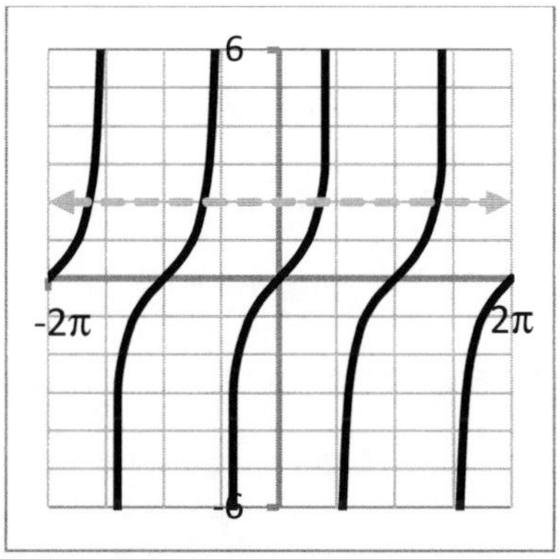

To make the function invertible, restrict the domain so that the truncated function does pass the horizontal line test. The only answer choice that accomplishes this is choice B: if you restrict the domain to $-\frac{\pi}{2} < x < \frac{\pi}{2}$, the resulting function will pass the horizontal line test as shown in the graph below.

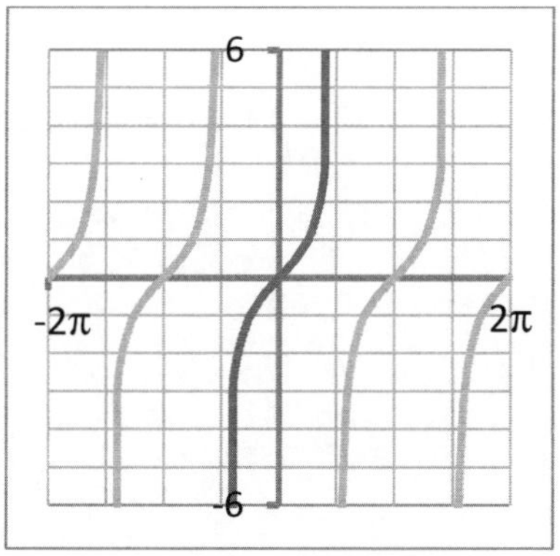

Math.Content.F.TF.7

47. B: First set up an equation for when the height of the weight is 15 inches:

$$h(t) = 15$$
$$20cos\ t + 30 = 15$$

To solve the equation, isolate cos *t*:

$$20cos\ t = -15$$
$$\cos t = -\frac{3}{4}$$

Then take the inverse cosine of both sides of the equation:

$$\cos^{-1}(\cos\ t) = \cos^{-1}\left(-\frac{3}{4}\right)$$

$$t = \cos^{-1}\left(-\frac{3}{4}\right)$$

Finally, use a calculator to calculate t and round your answer to the nearest tenth:

$$t \approx 2.4$$

Therefore, the weight is 15 inches above the ground after about 2.4 seconds.

Math.Content.F.TF.8

48. B: The Pythagorean identity is shown below:

$$\sin^2\theta + \cos^2\theta = 1$$

To make it easier to use in this problem, an equivalent identity can be written by subtracting $\cos^2\theta$ from both sides:

$$\sin^2\theta = 1 - \cos^2\theta$$

Substitute $\frac{12}{13}$ for $\cos\theta$ into this equation and solve for $\sin^2\theta$:

$$\begin{aligned}\sin^2\theta &= 1 - \cos^2\theta \\ &= 1 - \left(\frac{12}{13}\right)^2 \\ &= 1 - \frac{144}{169} \\ &= \frac{25}{169}\end{aligned}$$

Next take the square root of each side. Since θ is in quadrant IV, $\sin\theta$ is negative:

$$\sin^2\theta = \frac{25}{169}$$

$$\sin\theta = -\frac{5}{13}$$

You might also have memorized the Pythagorean triple {5, 12, 13} as the sides of a right triangle. If the cosine of an angle in the fourth quadrant is 12/13, then the side of the right triangle opposite that angle is 5 units long. The y-component of this point is negative, and so the sine, by SOHCAHTOA, is equal to -5/13.

Math.Content.F.TF.9

49. B: The addition formula for cosine is shown below:

$$\cos(\alpha + \beta) = \cos\alpha\cos\beta - \sin\alpha\sin\beta$$

Since you know the values of sin θ and cos θ when $\theta = \frac{\pi}{3}$ and when $\theta = \frac{\pi}{4}$, substitute these values for α and β into the addition formula. Then simplify the result:

$$\cos\left(\frac{7\pi}{12}\right) = \cos\left(\frac{\pi}{3} + \frac{\pi}{4}\right) = \cos\frac{\pi}{3}\cos\frac{\pi}{4} - \sin\frac{\pi}{3}\sin\frac{\pi}{4}$$

$$\begin{aligned}\cos\frac{7\pi}{12} &= \frac{1}{2}\cdot\frac{\sqrt{2}}{2} - \frac{\sqrt{3}}{2}\cdot\frac{\sqrt{2}}{2} \\ &= \frac{\sqrt{2}}{4} - \frac{\sqrt{6}}{4} \\ &= \frac{\sqrt{2} - \sqrt{6}}{4}\end{aligned}$$

Math.Content.F.TF.9

50. D: The addition formula for tangent is shown below:

$$\tan(\alpha + \beta) = \frac{\tan\alpha + \tan\beta}{1 - \tan\alpha\tan\beta}$$

To calculate tan $(x + y)$, substitute the values of tan x and tan y into the formula. Then simplify the result:

$$\begin{aligned}\tan(x + y) &= \frac{\tan x + \tan y}{1 - \tan x \tan y} \\ &= \frac{2/3 + 1/4}{1 - (2/3)(1/4)} \\ &= \frac{11/12}{1 - (1/6)} \\ &= \frac{11/12}{5/6} \\ &= \frac{11}{10}\end{aligned}$$

How to Overcome Test Anxiety

Just the thought of taking a test is enough to make most people a little nervous. A test is an important event that can have a long-term impact on your future, so it's important to take it seriously and it's natural to feel anxious about performing well. But just because anxiety is normal, that doesn't mean that it's helpful in test taking, or that you should simply accept it as part of your life. Anxiety can have a variety of effects. These effects can be mild, like making you feel slightly nervous, or severe, like blocking your ability to focus or remember even a simple detail.

If you experience test anxiety—whether severe or mild—it's important to know how to beat it. To discover this, first you need to understand what causes test anxiety.

Causes of Test Anxiety

While we often think of anxiety as an uncontrollable emotional state, it can actually be caused by simple, practical things. One of the most common causes of test anxiety is that a person does not feel adequately prepared for their test. This feeling can be the result of many different issues such as poor study habits or lack of organization, but the most common culprit is time management. Starting to study too late, failing to organize your study time to cover all of the material, or being distracted while you study will mean that you're not well prepared for the test. This may lead to cramming the night before, which will cause you to be physically and mentally exhausted for the test. Poor time management also contributes to feelings of stress, fear, and hopelessness as you realize you are not well prepared but don't know what to do about it.

Other times, test anxiety is not related to your preparation for the test but comes from unresolved fear. This may be a past failure on a test, or poor performance on tests in general. It may come from comparing yourself to others who seem to be performing better or from the stress of living up to expectations. Anxiety may be driven by fears of the future—how failure on this test would affect your educational and career goals. These fears are often completely irrational, but they can still negatively impact your test performance.

Elements of Test Anxiety

As mentioned earlier, test anxiety is considered to be an emotional state, but it has physical and mental components as well. Sometimes you may not even realize that you are suffering from test anxiety until you notice the physical symptoms. These can include trembling hands, rapid heartbeat, sweating, nausea, and tense muscles. Extreme anxiety may lead to fainting or vomiting. Obviously, any of these symptoms can have a negative impact on testing. It is important to recognize them as soon as they begin to occur so that you can address the problem before it damages your performance.

The mental components of test anxiety include trouble focusing and inability to remember learned information. During a test, your mind is on high alert, which can help you recall information and stay focused for an extended period of time. However, anxiety interferes with your mind's natural processes, causing you to blank out, even on the questions you know well. The strain of testing during anxiety makes it difficult to stay focused, especially on a test that may take several hours. Extreme anxiety can take a huge mental toll, making it difficult not only to recall test information but even to understand the test questions or pull your thoughts together.

Effects of Test Anxiety

Test anxiety is like a disease—if left untreated, it will get progressively worse. Anxiety leads to poor performance, and this reinforces the feelings of fear and failure, which in turn lead to poor performances on subsequent tests. It can grow from a mild nervousness to a crippling condition. If allowed to progress, test anxiety can have a big impact on your schooling, and consequently on your future.

Test anxiety can spread to other parts of your life. Anxiety on tests can become anxiety in any stressful situation, and blanking on a test can turn into panicking in a job situation. But fortunately, you don't have to let anxiety rule your testing and determine your grades. There are a number of relatively simple steps you can take to move past anxiety and function normally on a test and in the rest of life.

Physical Steps for Beating Test Anxiety

While test anxiety is a serious problem, the good news is that it can be overcome. It doesn't have to control your ability to think and remember information. While it may take time, you can begin taking steps today to beat anxiety.

Just as your first hint that you may be struggling with anxiety comes from the physical symptoms, the first step to treating it is also physical. Rest is crucial for having a clear, strong mind. If you are tired, it is much easier to give in to anxiety. But if you establish good sleep habits, your body and mind will be ready to perform optimally, without the strain of exhaustion. Additionally, sleeping well helps you to retain information better, so you're more likely to recall the answers when you see the test questions.

Getting good sleep means more than going to bed on time. It's important to allow your brain time to relax. Take study breaks from time to time so it doesn't get overworked, and don't study right before bed. Take time to rest your mind before trying to rest your body, or you may find it difficult to fall asleep.

Along with sleep, other aspects of physical health are important in preparing for a test. Good nutrition is vital for good brain function. Sugary foods and drinks may give a burst of energy but this burst is followed by a crash, both physically and emotionally. Instead, fuel your body with protein and vitamin-rich foods.

Also, drink plenty of water. Dehydration can lead to headaches and exhaustion, especially if your brain is already under stress from the rigors of the test. Particularly if your test is a long one, drink water during the breaks. And if possible, take an energy-boosting snack to eat between sections.

Along with sleep and diet, a third important part of physical health is exercise. Maintaining a steady workout schedule is helpful, but even taking 5-minute study breaks to walk can help get your blood pumping faster and clear your head. Exercise also releases endorphins, which contribute to a positive feeling and can help combat test anxiety.

When you nurture your physical health, you are also contributing to your mental health. If your body is healthy, your mind is much more likely to be healthy as well. So take time to rest, nourish your body with healthy food and water, and get moving as much as possible. Taking these physical steps will make you stronger and more able to take the mental steps necessary to overcome test anxiety.

Mental Steps for Beating Test Anxiety

Working on the mental side of test anxiety can be more challenging, but as with the physical side, there are clear steps you can take to overcome it. As mentioned earlier, test anxiety often stems from lack of preparation, so the obvious solution is to prepare for the test. Effective studying may be the most important weapon you have for beating test anxiety, but you can and should employ several other mental tools to combat fear.

First, boost your confidence by reminding yourself of past success—tests or projects that you aced. If you're putting as much effort into preparing for this test as you did for those, there's no reason you should expect to fail here. Work hard to prepare; then trust your preparation.

Second, surround yourself with encouraging people. It can be helpful to find a study group, but be sure that the people you're around will encourage a positive attitude. If you spend time with others who are anxious or cynical, this will only contribute to your own anxiety. Look for others who are motivated to study hard from a desire to succeed, not from a fear of failure.

Third, reward yourself. A test is physically and mentally tiring, even without anxiety, and it can be helpful to have something to look forward to. Plan an activity following the test, regardless of the outcome, such as going to a movie or getting ice cream.

When you are taking the test, if you find yourself beginning to feel anxious, remind yourself that you know the material. Visualize successfully completing the test. Then take a few deep, relaxing breaths and return to it. Work through the questions carefully but with confidence, knowing that you are capable of succeeding.

Developing a healthy mental approach to test taking will also aid in other areas of life. Test anxiety affects more than just the actual test—it can be damaging to your mental health and even contribute to depression. It's important to beat test anxiety before it becomes a problem for more than testing.

Study Strategy

Being prepared for the test is necessary to combat anxiety, but what does being prepared look like? You may study for hours on end and still not feel prepared. What you need is a strategy for test prep. The next few pages outline our recommended steps to help you plan out and conquer the challenge of preparation.

Step 1: Scope Out the Test

Learn everything you can about the format (multiple choice, essay, etc.) and what will be on the test. Gather any study materials, course outlines, or sample exams that may be available. Not only will this help you to prepare, but knowing what to expect can help to alleviate test anxiety.

Step 2: Map Out the Material

Look through the textbook or study guide and make note of how many chapters or sections it has. Then divide these over the time you have. For example, if a book has 15 chapters and you have five days to study, you need to cover three chapters each day. Even better, if you have the time, leave an extra day at the end for overall review after you have gone through the material in depth.

If time is limited, you may need to prioritize the material. Look through it and make note of which sections you think you already have a good grasp on, and which need review. While you are studying, skim quickly through the familiar sections and take more time on the challenging parts.

Write out your plan so you don't get lost as you go. Having a written plan also helps you feel more in control of the study, so anxiety is less likely to arise from feeling overwhelmed at the amount to cover.

Step 3: Gather Your Tools

Decide what study method works best for you. Do you prefer to highlight in the book as you study and then go back over the highlighted portions? Or do you type out notes of the important information? Or is it helpful to make flashcards that you can carry with you? Assemble the pens, index cards, highlighters, post-it notes, and any other materials you may need so you won't be distracted by getting up to find things while you study.

If you're having a hard time retaining the information or organizing your notes, experiment with different methods. For example, try color-coding by subject with colored pens, highlighters, or post-it notes. If you learn better by hearing, try recording yourself reading your notes so you can listen while in the car, working out, or simply sitting at your desk. Ask a friend to quiz you from your flashcards, or try teaching someone the material to solidify it in your mind.

Step 4: Create Your Environment

It's important to avoid distractions while you study. This includes both the obvious distractions like visitors and the subtle distractions like an uncomfortable chair (or a too-comfortable couch that makes you want to fall asleep). Set up the best study environment possible: good lighting and a comfortable work area. If background music helps you focus, you may want to turn it on, but otherwise keep the room quiet. If you are using a computer to take notes, be sure you don't have any other windows open, especially applications like social media, games, or anything else that could distract you. Silence your phone and turn off notifications. Be sure to keep water close by so you stay hydrated while you study (but avoid unhealthy drinks and snacks).

Also, take into account the best time of day to study. Are you freshest first thing in the morning? Try to set aside some time then to work through the material. Is your mind clearer in the afternoon or evening? Schedule your study session then. Another method is to study at the same time of day that you will take the test, so that your brain gets used to working on the material at that time and will be ready to focus at test time.

Step 5: Study!

Once you have done all the study preparation, it's time to settle into the actual studying. Sit down, take a few moments to settle your mind so you can focus, and begin to follow your study plan. Don't give in to distractions or let yourself procrastinate. This is your time to prepare so you'll be ready to fearlessly approach the test. Make the most of the time and stay focused.

Of course, you don't want to burn out. If you study too long you may find that you're not retaining the information very well. Take regular study breaks. For example, taking five minutes out of every hour to walk briskly, breathing deeply and swinging your arms, can help your mind stay fresh.

As you get to the end of each chapter or section, it's a good idea to do a quick review. Remind yourself of what you learned and work on any difficult parts. When you feel that you've mastered the material, move on to the next part. At the end of your study session, briefly skim through your notes again.

But while review is helpful, cramming last minute is NOT. If at all possible, work ahead so that you won't need to fit all your study into the last day. Cramming overloads your brain with more information than it can process and retain, and your tired mind may struggle to recall even previously learned information when it is overwhelmed with last-minute study. Also, the urgent nature of cramming and the stress placed on your brain contribute to anxiety. You'll be more likely to go to the test feeling unprepared and having trouble thinking clearly.

So don't cram, and don't stay up late before the test, even just to review your notes at a leisurely pace. Your brain needs rest more than it needs to go over the information again. In fact, plan to finish your studies by noon or early afternoon the day before the test. Give your brain the rest of the day to relax or focus on other things, and get a good night's sleep. Then you will be fresh for the test and better able to recall what you've studied.

Step 6: Take a Practice Test

Many courses offer sample tests, either online or in the study materials. This is an excellent resource to check whether you have mastered the material, as well as to prepare for the test format and environment.

Check the test format ahead of time: the number of questions, the type (multiple choice, free response, etc.), and the time limit. Then create a plan for working through them. For example, if you have 30 minutes to take a 60-question test, your limit is 30 seconds per question. Spend less time on the questions you know well so that you can take more time on the difficult ones.

If you have time to take several practice tests, take the first one open book, with no time limit. Work through the questions at your own pace and make sure you fully understand them. Gradually work up to taking a test under test conditions: sit at a desk with all study materials put away and set a timer. Pace yourself to make sure you finish the test with time to spare and go back to check your answers if you have time.

After each test, check your answers. On the questions you missed, be sure you understand why you missed them. Did you misread the question (tests can use tricky wording)? Did you forget the information? Or was it something you hadn't learned? Go back and study any shaky areas that the practice tests reveal.

Taking these tests not only helps with your grade, but also aids in combating test anxiety. If you're already used to the test conditions, you're less likely to worry about it, and working through tests until you're scoring well gives you a confidence boost. Go through the practice tests until you feel comfortable, and then you can go into the test knowing that you're ready for it.

Test Tips

On test day, you should be confident, knowing that you've prepared well and are ready to answer the questions. But aside from preparation, there are several test day strategies you can employ to maximize your performance.

First, as stated before, get a good night's sleep the night before the test (and for several nights before that, if possible). Go into the test with a fresh, alert mind rather than staying up late to study.

Try not to change too much about your normal routine on the day of the test. It's important to eat a nutritious breakfast, but if you normally don't eat breakfast at all, consider eating just a protein bar. If you're a coffee drinker, go ahead and have your normal coffee. Just make sure you time it so that the caffeine doesn't wear off right in the middle of your test. Avoid sugary beverages, and drink enough water to stay hydrated but not so much that you need a restroom break 10 minutes into the test. If your test isn't first thing in the morning, consider going for a walk or doing a light workout before the test to get your blood flowing.

Allow yourself enough time to get ready, and leave for the test with plenty of time to spare so you won't have the anxiety of scrambling to arrive in time. Another reason to be early is to select a good seat. It's helpful to sit away from doors and windows, which can be distracting. Find a good seat, get out your supplies, and settle your mind before the test begins.

When the test begins, start by going over the instructions carefully, even if you already know what to expect. Make sure you avoid any careless mistakes by following the directions.

Then begin working through the questions, pacing yourself as you've practiced. If you're not sure on an answer, don't spend too much time on it, and don't let it shake your confidence. Either skip it and come back later, or eliminate as many wrong answers as possible and guess among the remaining ones. Don't dwell on these questions as you continue—put them out of your mind and focus on what lies ahead.

Be sure to read all of the answer choices, even if you're sure the first one is the right answer. Sometimes you'll find a better one if you keep reading. But don't second-guess yourself if you do immediately know the answer. Your gut instinct is usually right. Don't let test anxiety rob you of the information you know.

If you have time at the end of the test (and if the test format allows), go back and review your answers. Be cautious about changing any, since your first instinct tends to be correct, but make sure you didn't misread any of the questions or accidentally mark the wrong answer choice. Look over any you skipped and make an educated guess.

At the end, leave the test feeling confident. You've done your best, so don't waste time worrying about your performance or wishing you could change anything. Instead, celebrate the successful completion of this test. And finally, use this test to learn how to deal with anxiety even better next time.

Review Video: Test Anxiety
Visit mometrix.com/academy and enter code: 100340

Important Qualification

Not all anxiety is created equal. If your test anxiety is causing major issues in your life beyond the classroom or testing center, or if you are experiencing troubling physical symptoms related to your anxiety, it may be a sign of a serious physiological or psychological condition. If this sounds like your situation, we strongly encourage you to seek professional help.

How to Overcome Your Fear of Math

Not again. You're sitting in math class, look down at your test, and immediately start to panic. Your stomach is in knots, your heart is racing, and you break out in a cold sweat. You're staring at the paper, but everything looks like it's written in a foreign language. Even though you studied, you're blanking out on how to begin solving these problems.

Does this sound familiar? If so, then you're not alone! You may be like millions of other people who experience math anxiety. Anxiety about performing well in math is a common experience for students of all ages. In this article, we'll discuss what math anxiety is, common misconceptions about learning math, and tips and strategies for overcoming math anxiety.

What Is Math Anxiety?

Psychologist Mark H. Ashcraft explains math anxiety as a feeling of tension, apprehension, or fear that interferes with math performance. Having math anxiety negatively impacts people's beliefs about themselves and what they can achieve. It hinders achievement within the math classroom and affects the successful application of mathematics in the real world.

Symptoms and Signs of Math Anxiety

To overcome math anxiety, you must recognize its symptoms. Becoming aware of the signs of math anxiety is the first step in addressing and resolving these fears.

Negative Self-Talk

If you have math anxiety, you've most likely said at least one of these statements to yourself:

- "I hate math."
- "I'm not good at math."
- "I'm not a math person."

The way we speak to ourselves and think about ourselves matters. Our thoughts become our words, our words become our actions, and our actions become our habits. Thinking negatively about math creates a self-fulfilling prophecy. In other words, if you take an idea as a fact, then it will come true because your behaviors will align to match it.

Avoidance

Some people who are fearful or anxious about math will tend to avoid it altogether. Avoidance can manifest in the following ways:

- Lack of engagement with math content
- Not completing homework and other assignments
- Not asking for help when needed
- Skipping class
- Avoiding math-related courses and activities

Avoidance is one of the most harmful impacts of math anxiety. If you steer clear of math at all costs, then you can't set yourself up for the success you deserve.

Lack of Motivation

Students with math anxiety may experience a lack of motivation. They may struggle to find the incentive to get engaged with what they view as a frightening subject. These students are often overwhelmed, making it difficult for them to complete or even start math assignments.

Procrastination

Another symptom of math anxiety is procrastination. Students may voluntarily delay or postpone their classwork and assignments, even if they know there will be a negative consequence for doing so. Additionally, they may choose to wait until the last minute to start projects and homework, even when they know they need more time to put forth their best effort.

Physiological Reactions

Many people with a fear of math experience physiological side effects. These may include an increase in heart rate, sweatiness, shakiness, nausea, and irregular breathing. These symptoms make it difficult to focus on the math content, causing the student even more stress and fear.

Strong Emotional Responses

Math anxiety also affects people on an emotional level. Responding to math content with strong emotions such as panic, anger, or despair can be a sign of math anxiety.

Low Test Scores and Performance

Low achievement can be both a symptom and a cause of math anxiety. When someone does not take the steps needed to perform well on tests and assessments, they are less likely to pass. The more they perform poorly, the more they accept this poor performance as a fact that can't be changed.

Feeling Alone

People who experience math anxiety feel like they are the only ones struggling, even if the math they are working on is challenging to many people. Feeling isolated in what they perceive as failure can trigger tension or nervousness.

Feeling of Permanency

Math anxiety can feel very permanent. You may assume that you are naturally bad at math and always will be. Viewing math as a natural ability rather than a skill that can be learned causes people to believe that nothing will help them improve. They take their current math abilities as fact and assume that they can't be changed. As a result, they give up, stop trying to improve, and avoid engaging with math altogether.

Lack of Confidence

People with low self-confidence in math tend to feel awkward and incompetent when asked to solve a math problem. They don't feel comfortable taking chances or risks when problem-solving because they second-guess themselves and assume they are incorrect. They don't trust in their ability to learn the content and solve problems correctly.

Panic

A general sense of unexplained panic is also a sign of math anxiety. You may feel a sudden sense of fear that triggers physical reactions, even when there is no apparent reason for such a response.

Causes Of Math Anxiety

Math anxiety can start at a young age and may have one or more underlying causes. Common causes of math anxiety include the following:

The Attitude of Parents or Guardians

Parents often put pressure on their children to perform well in school. Although their intentions are usually good, this pressure can lead to anxiety, especially if the student is struggling with a subject or class.

Perhaps your parents or others in your life hold negative predispositions about math based on their own experiences. For instance, if your mother once claimed she was not good at math, then you might have incorrectly interpreted this as a predisposed trait that was passed down to you.

Teacher Influence

Students often pick up on their teachers' attitudes about the content being taught. If a teacher is happy and excited about math, students are more likely to mirror these emotions. However, if a teacher lacks enthusiasm or genuine interest, then students are more inclined to disengage.

Teachers have a responsibility to cultivate a welcoming classroom culture that is accepting of mistakes. When teachers blame students for not understanding a concept, they create a hostile classroom environment where mistakes are not tolerated. This tension increases student stress and anxiety, creating conditions that are not conducive to inquiry and learning. Instead, when teachers normalize mistakes as a natural part of the problem-solving process, they give their students the freedom to explore and grapple with the math content. In such an environment, students feel comfortable taking chances because they are not afraid of being wrong.

Students need teachers that can help when they're having problems understanding difficult concepts. In doing so, educators may need to change how they teach the content. Since different people have unique learning styles, it's the job of the teacher to adapt to the needs of each student. Additionally, teachers should encourage students to explore alternate problem-solving strategies, even if it's not the preferred method of the educator.

Fear of Being Wrong

Embarrassing situations can be traumatic, especially for young children and adolescents. These experiences can stay with people through their adult lives. Those with math anxiety may experience a fear of being wrong, especially in front of a group of peers. This fear can be paralyzing, interfering with the student's concentration and ability to focus on the problem at hand.

Timed Assessments

Timed assessments can help improve math fluency, but they often create unnecessary pressure for students to complete an unrealistic number of problems within a specified timeframe. Many studies have shown that timed assessments often result in increased levels of anxiety, reducing a student's overall competence and ability to problem-solve.

Debunking Math Myths

There are lots of myths about math that are related to the causes and development of math-related anxiety. Although these myths have been proven to be false, many people take them as fact. Let's go over a few of the most common myths about learning math.

Myth: Men Are Better at Math Than Women

Math has a reputation for being a male-dominant subject, but this doesn't mean that men are inherently better at math than women. Many famous mathematical discoveries have been made by women. Katherine Johnson, Dame Mary Lucy Cartwright, and Marjorie Lee Brown are just a few of the many famous women mathematicians. Expecting to be good or bad at math because of your gender sets you up for stress and confusion. Math is a skill that can be learned, just like cooking or riding a bike.

Myth: There Is Only One Good Way to Solve Math Problems

There are many ways to get the correct answer when it comes to math. No two people have the same brain, so everyone takes a slightly different approach to problem-solving. Moreover, there isn't one way of problem-solving that's superior to another. Your way of working through a problem might differ from someone else's, and that is okay. Math can be a highly individualized process, so the best method for you should be the one that makes you feel the most comfortable and makes the most sense to you.

Myth: Math Requires a Good Memory

For many years, mathematics was taught through memorization. However, learning in such a way hinders the development of critical thinking and conceptual understanding. These skill sets are much more valuable than basic memorization. For instance, you might be great at memorizing mathematical formulas, but if you don't understand what they mean, then you can't apply them to different scenarios in the real world. When a student is working from memory, they are limited in the strategies available to them to problem-solve. In other words, they assume there is only one correct way to do the math, which is the method they memorized. Having a variety of problem-solving options can help students figure out which method works best for them. Additionally, it provides students with a better understanding of how and why certain mathematical strategies work. While memorization can be helpful in some instances, it is not an absolute requirement for mathematicians.

Myth: Math Is Not Creative

Math requires imagination and intuition. Contrary to popular belief, it is a highly creative field. Mathematical creativity can help in developing new ways to think about and solve problems. Many people incorrectly assume that all things are either creative or analytical. However, this black-and-white view is limiting because the field of mathematics involves both creativity and logic.

Myth: Math Isn't Supposed To Be Fun

Whoever told you that math isn't supposed to be fun is a liar. There are tons of math-based activities and games that foster friendly competition and engagement. Math is often best learned through play, and lots of mobile apps and computer games exemplify this.

Additionally, math can be an exceptionally collaborative and social experience. Studying or working through problems with a friend often makes the process a lot more fun. The excitement and satisfaction of solving a difficult problem with others is quite rewarding. Math can be fun if you look for ways to make it more collaborative and enjoyable.

MYTH: NOT EVERYONE IS CAPABLE OF LEARNING MATH

There's no such thing as a "math person." Although many people think that you're either good at math or you're not, this is simply not true. Everyone is capable of learning and applying mathematics. However, not everyone learns the same way. Since each person has a different learning style, the trick is to find the strategies and learning tools that work best for you. Some people learn best through hands-on experiences, and others find success through the use of visual aids. Others are auditory learners and learn best by hearing and listening. When people are overwhelmed or feel that math is too hard, it's often because they haven't found the learning strategy that works best for them.

MYTH: GOOD MATHEMATICIANS WORK QUICKLY AND NEVER MAKE MISTAKES

There is no prize for finishing first in math. It's not a race, and speed isn't a measure of your ability. Good mathematicians take their time to ensure their work is accurate. As you gain more experience and practice, you will naturally become faster and more confident.

Additionally, everyone makes mistakes, including good mathematicians. Mistakes are a normal part of the problem-solving process, and they're not a bad thing. The important thing is that we take the time to learn from our mistakes, understand where our misconceptions are, and move forward.

MYTH: YOU DON'T NEED MATH IN THE REAL WORLD

Our day-to-day lives are so infused with mathematical concepts that we often don't even realize when we're using math in the real world. In fact, most people tend to underestimate how much we do math in our everyday lives. It's involved in an enormous variety of daily activities such as shopping, baking, finances, and gardening, as well as in many careers, including architecture, nursing, design, and sales.

Tips and Strategies for Overcoming Math Anxiety

If your anxiety is getting in the way of your level of mathematical engagement, then there are lots of steps you can take. Check out the strategies below to start building confidence in math today.

FOCUS ON UNDERSTANDING, NOT MEMORIZATION

Don't drive yourself crazy trying to memorize every single formula or mathematical process. Instead, shift your attention to understanding concepts. Those who prioritize memorization over conceptual understanding tend to have lower achievement levels in math. Students who memorize may be able to complete some math, but they don't understand the process well enough to apply it to different situations. Memorization comes with time and practice, but it won't help alleviate math anxiety. On the other hand, conceptual understanding will give you the building blocks of knowledge you need to build up your confidence.

Replace Negative Self-Talk with Positive Self-Talk

Start to notice how you think about yourself. Whenever you catch yourself thinking something negative, try replacing that thought with a positive affirmation. Instead of continuing the negative thought, pause to reframe the situation. For ideas on how to get started, take a look at the table below:

Instead of thinking...	Try thinking...
"I can't do this math." "I'm not a math person."	"I'm up for the challenge, and I'm training my brain in math."
"This problem is too hard."	"This problem is hard, so this might take some time and effort. I know I can do this."
"I give up."	"What strategies can help me solve this problem?"
"I made a mistake, so I'm not good at this."	"Everyone makes mistakes. Mistakes help me to grow and understand."
"I'll never be smart enough."	"I can figure this out, and I am smart enough."

Practice Mindfulness

Practicing mindfulness and focusing on your breathing can help alleviate some of the physical symptoms of math anxiety. By taking deep breaths, you can remind your nervous system that you are not in immediate danger. Doing so will reduce your heart rate and help with any irregular breathing or shakiness. Taking the edge off of the physiological effects of anxiety will clear your mind, allowing your brain to focus its energy on problem-solving.

Do Some Math Every Day

Think about learning math as if you were learning a foreign language. If you don't use it, you lose it. If you don't practice your math skills regularly, you'll have a harder time achieving comprehension and fluency. Set some amount of time aside each day, even if it's just for a few minutes, to practice. It might take some discipline to build a habit around this, but doing so will help increase your mathematical self-assurance.

Use All of Your Resources

Everyone has a different learning style, and there are plenty of resources out there to support all learners. When you get stuck on a math problem, think about the tools you have access to, and use them when applicable. Such resources may include flashcards, graphic organizers, study guides, interactive notebooks, and peer study groups. All of these are great tools to accommodate your individual learning style. Finding the tools and resources that work for your learning style will give you the confidence you need to succeed.

Realize That You Aren't Alone

Remind yourself that lots of other people struggle with math anxiety, including teachers, nurses, and even successful mathematicians. You aren't the only one who panics when faced with a new or challenging problem. It's probably much more common than you think. Realizing that you aren't alone in your experience can help put some distance between yourself and the emotions you feel about math. It also helps to normalize the anxiety and shift your perspective.

Ask Questions

If there's a concept you don't understand and you've tried everything you can, then it's okay to ask for help! You can always ask your teacher or professor for help. If you're not learning math in a traditional classroom, you may want to join a study group, work with a tutor, or talk to your friends. More often than not, you aren't the only one of your peers who needs clarity on a mathematical concept. Seeking understanding is a great way to increase self-confidence in math.

Remember That There's More Than One Way To Solve A Problem

Since everyone learns differently, it's best to focus on understanding a math problem with an approach that makes sense to you. If the way it's being taught is confusing to you, don't give up. Instead, work to understand the problem using a different technique. There's almost always more than one problem-solving method when it comes to math. Don't get stressed if one of them doesn't make sense to you. Instead, shift your focus to what does make sense. Chances are high that you know more than you think you do.

Visualization

Visualization is the process of creating images in your mind's eye. Picture yourself as a successful, confident mathematician. Think about how you would feel and how you would behave. What would your work area look like? How would you organize your belongings? The more you focus on something, the more likely you are to achieve it. Visualizing teaches your brain that you can achieve whatever it is that you want. Thinking about success in mathematics will lead to acting like a successful mathematician. This, in turn, leads to actual success.

Focus on the Easiest Problems First

To increase your confidence when working on a math test or assignment, try solving the easiest problems first. Doing so will remind you that you are successful in math and that you do have what it takes. This process will increase your belief in yourself, giving you the confidence you need to tackle more complex problems.

Find a Support Group

A study buddy, tutor, or peer group can go a long way in decreasing math-related anxiety. Such support systems offer lots of benefits, including a safe place to ask questions, additional practice with mathematical concepts, and an understanding of other problem-solving explanations that may work better for you. Equipping yourself with a support group is one of the fastest ways to eliminate math anxiety.

Reward Yourself for Working Hard

Recognize the amount of effort you're putting in to overcome your math anxiety. It's not an easy task, so you deserve acknowledgement. Surround yourself with people who will provide you with the positive reinforcement you deserve.

Remember, You Can Do This!

Conquering a fear of math can be challenging, but there are lots of strategies that can help you out. Your own beliefs about your mathematical capabilities can limit your potential. Working toward a growth mindset can have a tremendous impact on decreasing math-related anxiety and building confidence. By knowing the symptoms of math anxiety and recognizing common misconceptions about learning math, you can develop a plan to address your fear of math. Utilizing the strategies discussed can help you overcome this anxiety and build the confidence you need to succeed.

Thank You

We at Mometrix would like to extend our heartfelt thanks to you, our friend and patron, for allowing us to play a part in your journey. It is a privilege to serve people from all walks of life who are unified in their commitment to building the best future they can for themselves.

The preparation you devote to these important testing milestones may be the most valuable educational opportunity you have for making a real difference in your life. We encourage you to put your heart into it—that feeling of succeeding, overcoming, and yes, conquering will be well worth the hours you've invested.

We want to hear your story, your struggles and your successes, and if you see any opportunities for us to improve our materials so we can help others even more effectively in the future, please share that with us as well. **The team at Mometrix would be absolutely thrilled to hear from you!** So please, send us an email (support@mometrix.com) and let's stay in touch.

If you'd like some additional help, check out these other resources we offer for your exam: http://MometrixFlashcards.com/CCSS

Online Resources

Due to our efforts to try to keep this book to a manageable length, we've created a link that will give you access to all of your online resources:

mometrix.com/resources719/cchsfunctions

It's Your Moment, Let's Celebrate It!

Share your story @mometrixtestpreparation